Lyssandra Kelly Silva Ferreira
Fabiana B. Frazão
Elaine C. B.Santos

Yield and physico-chemical evaluation of Uritinga fillet

Lyssandra Kelly Silva Ferreira
Fabiana B. Frazão
Elaine C. B.Santos

Yield and physico-chemical evaluation of Uritinga fillet

Meat, carcass and waste yields Physico-chemical analysis of the fillet

ScienciaScripts

Imprint
Any brand names and product names mentioned in this book are subject to trademark, brand or patent protection and are trademarks or registered trademarks of their respective holders. The use of brand names, product names, common names, trade names, product descriptions etc. even without a particular marking in this work is in no way to be construed to mean that such names may be regarded as unrestricted in respect of trademark and brand protection legislation and could thus be used by anyone.

Cover image: www.ingimage.com

This book is a translation from the original published under ISBN 978-613-9-73566-2.

Publisher:
Sciencia Scripts
is a trademark of
Dodo Books Indian Ocean Ltd. and OmniScriptum S.R.L publishing group

120 High Road, East Finchley, London, N2 9ED, United Kingdom
Str. Armeneasca 28/1, office 1, Chisinau MD-2012, Republic of Moldova, Europe
Printed at: see last page
ISBN: 978-620-7-87208-4

I dedicate

To the most precious things that God has given me in my life, my parents, my sister and my friends, for giving me marvellous moments in my daily life.

ACKNOWLEDGEMENT

To my parents, Francisco de Paulo Ferreira Filho and Lezita Ferreira, and my sister, Lygyanne de Oliveira, for their financial encouragement to carry out this work, as well as their unconditional understanding so that I could achieve my goal exactly.

Prof[3] Elaine Cristina Batista dos Santos, for her guidance and dedication to you during the development of this work, which contributed to my professional training.

To Prof Dr Haroldo Gomes Barroso, for his determination and bravery in the face of the difficulties encountered during the development of the Fisheries Engineering course, as well as for his encouragement to pursue this chosen path with faith and hope.

To Prof[3] Erivânia Gomes, for her help during the development of the work, as well as the advice and knowledge she acquired during its preparation.

To Prof[3] Msc. Marina Bezerra Figueiredo for her advice and guidance on academic research.

To the members of the Maranhão Squid Mollusc Study Group - GELM_MA, Flávia Abreu, Fabiana Borralho, Tatiana Pereira, students on the Fisheries Engineering course at UEMA, for their understanding, patience, assiduous companionship and irreplaceable opportunities and moments during this journey.

I would therefore like to thank the people who have participated directly and indirectly in this academic journey and victory.

"Never walk away from your dreams, because if they're gone, you'll still be alive, but you'll have ceased to exist."

Charles Chaplin

Summary

SUMMARY

The aim of this study was to assess the morphometric characteristics, fillet yield and physico-chemical parameters of the uritinga *Sciades proops* (Valenciennes, 1840). Thirty specimens of the species were used, distributed in a completely causalised experimental design, with each specimen considered a replicate. Morphometric ratios were assessed, such as: head length/standard length (CC/CP), dorsal height/standard length (LD/CP) and dorsal height/head length (LD/CC), followed by fillet yield with and without skin, carcass yield with and without head and percentage of waste such as head (PCab) and viscera (PV). In the physico-chemical analysis, they used 30 skinless fillets with a 10-repetition design for each parameter. The parameters observed were moisture (UM), protein (PB), water loss due to defrosting and cooking, and water retention capacity (CRA). The morphometric ratios CC/CP, LD/CP and LD/CC were found to have average values of 25.25% (± 1.85), 16.10% (± 1.36) and 158.05% (± 12.00), respectively, which characterises the species as having an elongated body and a moderately large compressed head. With regard to the yields obtained, it was observed that the fillets with skin gave respective values of between 39.89 and 53.49% and 33.07 and 46.05%, showing that carrying out the skin removal process reduced the fillet yield by 6.84 to 7.34%, while the carcass yields with head and without head showed respective average values of 92.17% (±1.42) and 59.07% (±2.65), of which the head made up around 31.10 to 39.49%. With regard to the yield of waste generated, the head was found to have the highest average percentage, corresponding to around 33.10%. The physico-chemical parameters of the uritinga showed 78.45% MU, 17.74% PB, 7.5 pH, 5.41% water loss on defrosting, 21.22% water loss on cooking and 51.55% CRA. It can be concluded that the data on morphometrics, fillet yield and centesimal composition associated with physico-chemical parameters (water loss, CRA and pH) showed that the species has good potential for industrialisation. However, in order to use the species in the seafood industry, further studies are needed on when the species is most and least available, the best processing techniques that can be used, and verification of economic viability[1] , among other factors.

Keywords: morphometrics, bromatology, retail, São Luís.

CHAPTER 1

INTRODUCTION

Brazil has a coastal zone that includes a diversity of contiguous ecosystems, which favours the development of various economic activities such as extractive fishing, which is currently considered to be the main source of fish supply for Brazilian society, a fact observed in the Statistical Bulletin of Fisheries and Aquaculture - Brazil 2010, which reports that, in 2010, this type of fishing contributed around 62.09% of national fish production, of which 536,455 tonnes were obtained through marine extractive fishing (SOARES, 1999; CORSON; SERAFIM; HAZIM, 2009; MPA.2011), was obtained through marine extractive fishing (SOARES, 1999; CORSON, 2002; SERAFIM; HAZIN, 2006, KNOPPERS, 2009; MPA.2011).

In the panorama of extractive marine fishing in Brazil, the state of Maranhão is the fifth largest producer of fish with 43,780.1 tonnes (MPA, 2011). This productivity in Maranhão is due to the involvement of approximately 104.000 fishermen and the physiographic conditions found along the coastal zone, which favour the peculiar characteristics of the various existing ecosystems, such as the vast extension of the continental shelf, high primary production, which contribute to the development of biodiversity with a large representation of species of commercial value (SUDENE, 1983; STRIDE, 1992; CAMARGO; ISAAC, 2001; RIOS, 2005; IBAMA/SEAP/PROZEE 2008; ALMEIDA, 2008; FERREIRA et.aL, 2009, PEREIRA et aL, 2011).

Among the fishery resources exploited in Maranhão's maritime waters, the species *Sciades proops* (Valenciennes, 1840), commonly known as "uritinga", stands out. This is a native species that has a high catch rate throughout the state's coastline, with an average annual production of 1,305 tonnes. This shows its fishing potential due to its abundance, occurrence at all times of the year and ease of catching with any gear (STRIDE, 1992; RIOS, 2005; IABS, 2006; FERREIRA et al. 2011). According to Almeida (2008), this species is among the top ten species favoured by consumers in the state market.

The commercialisation of uritinga consists of negotiating a whole or gutted product through intermediary agents (middlemen, freight forwarders, balancers and pointers), and this product is then passed on to the retail trade such as street markets, fishmongers, markets and supermarkets.

According to Smith (1979), Falabella (1994), Evangelista (1998), Ostrensky et al. (2000), Caraciolo et al. (2001) and Bombardelli et al. (2005), the marketing of fishery products from native Brazilian species is generally limited to the simplest and least elaborate forms of processing, as well as a lack of standardisation and practicality, resulting from a series of factors such as: a lack of knowledge about the composition of nutrients and, above all, the "food value" of the product.

Emphasising that if these aspects are associated with sanitary and technological problems, resulting from deficiencies in handling, maintenance of the cold chain, conservation, storage, lack of adequate processing techniques and others, they have a direct influence on the consumer's choice of fish meat (GAGLEAZZI et al., 2002; OETTERER, 2002; TAVARES; GONÇALVES, 2011).

In order to overcome this paradigm and meet the growing demand from the consumer market for high quality products, the fish processing industries have been using traditional, innovative and emerging conservation technologies, the purpose of which is to offer society products from extractive fishing that are microbiologically safe, as well as providing minimal biochemical alterations, promoting the maintenance of the nutritional and sensory quality of the food and also prolonging their shelf life, making them marketable for much longer (LEISTNER; GORRIS, 1995; OLIVEIRA; GONÇALVES, 2011; NOVAES et aL,2012).

In addition, industrialised products must have characteristics that consumers want, such as good presentation (suitable cuts), convenience, easy preparation, hygiene, quality packaging and nutritional advantages (OETTERER, 2002; SOUZA, 2002).

It is also worth mentioning that the fish processing methods used today range from preparing fillets with or without skin, cooling, freezing, canning, drying, salting, smoking, fermentation, marinating, mechanically separated meat (MSM), surumi to obtaining breaded products, sausages and making use of by-products. These modernisation processes enable greater value to be added to fishery products and by-products, as well as allowing them to be popularised (OGAWA; MAIA, 1999; BOMBARDELLI et al, 2005; OLIVEIRA; GONÇALVES, 2011).

It is important to note that in order to subject fish to any processing method or techniques for industrialisation, data is required on the morphometric characteristics of the specific species, such as body size and weight, age, sex, anatomical shape, head size and weight of viscera, skin and fins (CONTRERAS-GUZMÁN, 1994; GOMIERO et aL, 2003).

According to Contreras-Guzmán (1994) and Gomiero et aL (2003), this knowledge of morphometric relationships served as a basis for obtaining meat cuts, yield data and the chemical composition of the meat, identifying the amount of edible part and nutrients that make up the fillet.

If this knowledge is combined with information on physical properties, such as pH, water retention capacity (WRC), a set of quality parameters is formed that guarantees the maintenance of meat tenderness and juiciness, the aesthetic appearance of the fish and other qualitative attributes of fish meat. These quality parameters play a decisive role when purchasing a fish product, especially those from marine extractive fisheries.

Meanwhile, there is a lack of studies on the morphometric characteristics, fillet yield and physical and bromatological composition of native fish from marine extractive fisheries. It is therefore necessary to carry out research into the morphometric characteristics, fillet yield and physico-chemical parameters of native species, the purpose of which is to provide information on the industrialisation potential of new species.

CHAPTER 2

OBJECTIVES

2.1 General

Evaluating the morphometric characteristics, fillet yield and physico-chemical parameters of the uritinga *Sciades proops* (Valenciennes, 1840) from the retail trade in the city of São Luís-MA.

2.2 Specific

- Describe the morphometric relationships used for processing species with industrialisation potential;
- To analyse the fillet yield of the species under study;
- To analyse the centesimal composition of the fillets through moisture and crude protein;
- Determine the physical parameters: pH, water retention capacity (WRC) and water loss due to defrosting and cooking; and
- To provide subsidies for the processing process.

CHAPTER 3

THEORETICAL BACKGROUND

Per capita consumption of fish products has shown continuous and vigorous growth in recent years, mainly due to their nutritional advantages for health, resulting from their functional components, especially their highly digestible protein content and high content of polyunsaturated fatty acids. In addition, they fulfil consumer requirements such as: diversity of products and by-products, price, product quality, convenience, easy preparation, good presentation, etc. (BRUSCHI, 2001; FAO, 2007; SHIRATO; SONADA, 2004; TRONDSEN et al, 2004).

Among the fishery products of greatest economic and culinary interest is the fillet, which consists of a semi-finished product made up mainly of the dorsal-lateral muscles of the fish, which may have a skin or no skin at all (WILLE et al, 2002, MACEDO-VIEGAS; SOUZA, 2004).

However, there is a lack of diversity in fish fillets on the domestic market, due to the absence of information on the influence of body characteristics on fillet and carcass yields, the existence of standardised filleting methods and physico-chemical properties for the majority of fish species, so there is a gap in data that allows species to be compared, critical factors to be assessed and the potential for industrialisation to be visualised (WILLE et aL, 2002, MACEDO-VIEGAS; SOUZA, 2004; SHIRATO; SONADA , 2004).

However, according to Valenti et al (2000), Wille et aL (2002) and Macedo-Viegas and Souza (2004), the importance of knowledge related to morphometric characteristics and fillet yield associated with physical and chemical properties has become an extremely important tool in defining the body characteristics that species with potential for industrialisation should have, the purpose of which is to provide greater meat utilisation, as well as guaranteeing the maintenance of nutritional values, the quality of the product handled and providing the necessary subsidies for the process of preserving and preparing the product.

3.1 Body morphometric characteristics

Morphometry is a quantitative study related to the variation and covariation of morphological measurements of organisms, whether these are pairs of anatomically homologous points, or between tangential parts or ends of structures (BOOKSTEIN et al. 1985; HUANG;LIAO, 1990; SWAIN; FOOTE, 1999; MORAES, 2003; ROCHA et al., 2009).

According to Blackith and Reyment (1971), Bosworth et al. (1998) and Cibert et al. (1999), morphometry is a tool used to describe the anatomical characteristics of a given species of fish, such as body shape, head size and weights of waste products (viscera, skin, fins, gonads and gills), as well as to quantify morphological differences between organisms.

Morphometric measurements vary according to the inter- and intraspecific characteristics of the species under study, as well as the action of exogenous factors such as climate, season, water quality, abundance and type of feed, which can affect the body's capacity to accumulate muscle mass and, consequently, the yield of edible parts (EYO, 1993; CONTRERAS-GUZMÁN, 1994; BOSWORTH et al.,1998).CIBERT et al, 1999; RIBEIRO et al, 1998; REIDEL et al, 2004; OLIVEIRA, 2012).

It is important to point out that morphometric ratios are considered important parameters for industrialising fish, especially head length/standard length (HC/LC), head length/head height (HC/ HC), standard length/total length (HC/TL) and total height/trunk length (TL/TL), which directly influence carcass and fillet yields (CONTRERAS-GUZMÁN, 1994; RIBEIRO et al., 1998; REIDEL et al., 2004).; OLIVEIRA, 2012).

Morphometric measurements are therefore a tool that can be used in studies to assess fillet and carcass yield and their correlation (GOODMAN, 1973; RUTTEN et al, 2004; FREATO et al, 2005; ROCHA et al, 2009).

Information on the use of morphometrics (weight, height, length, width and other dimensions of the fish) to determine the fillet and carcass yield of various fish species has already been verified in various studies such as:

- Miranda and Ribeiro (1997): who analysed the morphometric relationships of height/standard length , width/standard length and perimeter/standard length for determining fillet and carcass yields in surubim *(Pseudoplastiloma coruscans)* and observed that the coporal size of the species influenced the yields obtained.

- Bombardelli and Sanches (2008), who assessed morphometric ratios such as head length/standard length (HC/LC), head length/head height (HC/HH), among others, of the armado (*Pterodoras granulosas),* observed a reduction in the yield of the edible part of the fish and fillet yield when compared to pacu, tilapia, curimbatá and other species, due to the large dimensions of the species' head;

- Eyo (1993): reported a similar situation to that found by Bombardelli and Sanches (2008) in their study on the carcass composition and filleting yield of ten species of fish from Lake Kainji and corroborated that the processing yield is a reflection of the anatomical structure, i.e. fish with a large head in relation to the body had a lower fillet yield when compared to those with a small head;

- Santos et al (2006): carried out measurements of total length, standard length, total weight, fillet weight with and without leather and leather weight to determine the fillet yield of the palometa *(Serrasalmus spilopleura)* and observed that the total weight and total length of the palometa allows them to estimate the fillet weight obtained;
- Rutten et al. (2005): proposed in their study of genetic parameters for fillet characteristics and body measurements in nilotic tilapia *(Oreochromis niloticus)* that standard width and length can be used as determining requirements when verifying fillet weight and yield in nilotic tilapia;

■ Leonhardt et al. (2006): analysed the morphometric characteristics, yield and fillet composition of nile tilapia *(Oreochromis niloticus),* of the Thai chitralada strain and the strain from a cross between the two, and found that morphometric relationships, particularly head dimensions, influenced tilapia fillet yield;

■ Freato (2005): found that when evaluating carcasses of piracanjubas *(Brycon orbignyanus),* height and standard length are the most suitable criteria for selection;

■ Sang et al. (2009): observed in their study on the prediction of fillet weight and fillet yield for catfish *(Pangasianodon hypophthalmus),* that morphometrics can be used to predict fillet yield and body weight, which is an important practice for the processing industries; and

■ Other studies: observed the influence of morphometric measurements on fillet yield in common carp (CIBERT et al, 1999) and channel catfish (BOSWORTH et al, 2001).

In this context, knowledge of the anatomical shapes of species expressed through morphometrics is of great importance from an industrial point of view, as it provides basic information for choosing the appropriate techniques and equipment to be used in the processes of decapitation, evisceration and general cleaning of the raw material, as well as the sizing of boxes, chambers and other storage equipment; the process of obtaining good shapes (cuts) for presenting the product (steaks or fillets); the speed of cooling with ice and freezing to be used and targeting the ideal size and weight of the species that promotes the best carnea yield for the industry, meeting the demands of the most sophisticated consumer markets (EYO,1993; CONTRERAS-GUZMÁN,1994; RIBEIRO et aL, 1998; PINHEIRO et aL , 2006).

3.2 Fillet yield

Fish consumption in Brazil currently stands at around 9.03 Kg/inhabitant/year, but this is still below the World Health Organisation (WHO) recommendation of 12

Kg/inhabitant/year and the world average of 17 Kg per inhabitant (FAO, 2010; SIDONIO et al, 2012).

According to Coelho (2003) and Souza (2003), one of the ways of stimulating fish consumption in Brazilian society, and consequently solving the problem of low consumption, is to use a variety of processing techniques, ingredients and packaging to obtain products with nutritional and sensory quality and a good presentation.

Among the wide variety of fish-based products and by-products available for consumption, fillet stands out as the product of greatest interest to Brazilian society, as it has the characteristics demanded by consumers in relation to fish meat, such as: white meat with a firm texture, delicate flavour, absence of bones and an unpleasant odour (SOUZA, 2002; RIOS et aL, 2009).

To obtain fillets, the filleting technique is used, which is a form of fish processing that normally consists of the following stages (BOMBARDELLI et al, 2005; SARMIENTO, 2006; SILVA et al, 2008):

- *Raw material reception*

- *Stunning*

It is a procedure carried out on the raw material by means of electronarcosis or thermonarcosis, the purpose of which is to minimise physical damage and promote a longer period of *rigor mortis,* favouring the quality of the meat. Generally, stunned fish are placed in containers containing water and ice (PADUA, 2000).

- *Selection and classification*

The raw material is submitted to a selection process based on morphometric characteristics (ideal size and weight), state of health (absence of spots and wounds on the animal's body, intact fins, etc.) and others (CYRINO et al, 2004).

- *Evisceration*

The internal organs of the fish are removed, the purpose of which is to eliminate bacteria and digestive enzymes that could contaminate the fish's muscles.

This stage must be carried out quickly after the fish has died from evisceration (OETERRER, 1998).

- *Beheading*

It consists of removing the head and consequently removing the gills, which are considered to be one of the main sources of microorganisms (CYRINO et al., 2004).

- *Washing*

After eviscerating and decapitating the raw material, it is washed with chlorinated water (05 mg L'1 NaClO) to remove the remains of blood and mucus (CYRINO et al, 2004).

- *Filleting*

According to Souza and Macedo-Viegas (2000), this involves removing the dorsal and abdominal muscles from the fish using a knife as auxiliary equipment. At this stage, fillets are obtained, which may or may not have skin when they are marketed.

- *Packaging and storage*

The products obtained can be packed in plastic bags, expanded trays and cardboard boxes, then stored under refrigeration at between 0.5 °C and -2.0 °C (SARMIENTO, 2006).

This type of processing is generally used in fish species that do not have Y-shaped spines in their muscles, such as tilapia (*Oreochromis* ssp.)(BOMBARDELLI et al., 2005), surubim *(Pseudoplastiloma* coruscans)(CREPALDI et al, 2006), jundiás (*Rhamdia* qi/eZen)(LOPES et al, 2006), corvinas (*Micropogonias* furn/e/7)(GONÇALVES and PASSOS, 2010) and pirarucus *(Arapaima* gZgas)(NUNES, 2011).

However, there are species that have non-articulated spines on the vertebrae called intramuscular bones, which are inserted into the connective tissue of the muscle, attaching to the vertebral column, and these spines are small, straight or curved in large numbers. These characteristics are notorious in the curimbatá *(Prochilodus lineatus),* piavuçu (*Leporinas macrocephalus)* (OLIVEIRA et al., 2001; REIDEL et al.,2004), matrinxã (*Brycon cephalus)(GOM\ERO* et aL, 2003), pacu (*Piaractus mesopotamicus)* (BITTENCOURT, 2008) and tambaqui *(Colossoma macropomum)* (KUBITZA, 2004). Despite this peculiarity, they can still be subjected to the filleting process, as the presence of these spines is not a conditioning factor for filleting fish, but the fillets obtained had spines and therefore influenced their quality (OLIVEIRA et aL, 2001).

The filleting process is considered to be the main fish meat processing carried out in Brazil. However, there is still no standardisation for its execution, and it is done in different ways, leading to disagreement over the best method to use, i.e. choosing the method that provides the highest fillet yield, as well as operational ease and the shortest processing time (OETERRER, 1998; SOUZA; MACEDO-VIEGAS, 2000; SOUZA, 2002; BOMBARDELLI et al., 2005).

According to Oeterrer (1998) and Souza and Maranhão (2001), the standardisation of filleting techniques associated with the ideal size of the raw material, which allows the use of standard cuts, are criteria that need to be established in order to obtain higher fillet yields.

There is therefore a need to develop studies to assess fillet yields, as well as carcass yields (remaining meat + bone), waste (head, scales, viscera and skin), and the percentage of by-products that can be used for industrialisation, depending on the slaughter weight of the fish (SOUZA; MARANHÃO, 2001; COELHO, 2003).

Knowledge of the proportion of the raw material that will be transformed into final products for commercialisation, as well as the quality that will be part of the carcass and processing residues, provides the necessary information for logistical production planning and calculations for assessing the production efficiency of industries. In addition, this knowledge provides the industry with information on the potential for industrialisation and minimum processing of a given species (EYO, 1993; LANARI etaL, 1999; SOUZA et al, 1999; FARIA et al, 2003; LEONHARDT et al, 2006).

When carrying out these studies, some aspects must be taken into account that end up influencing processing yields, such as operator skill, filleting method, degree of mechanisation (filleting machines) and some intrinsic characteristics of the raw material, such as body shape, sex, size, age, and weight of viscera, skin and fins (CONTRERAS-GUZMÁN, 1994; SOUZA et al.,1998; MACEDO-VIEGAS et al, 2000; MACEDO-VIEGAS; SOUZA,2004).

In Brazil, studies related to fish carcass and fillet yields are scarce, and these present controversial data in relation to the standardisation of terms, causing great disparity in their use, with no consensus on their use, which makes it difficult to compare the parameters analysed in the different studies (MACEDO-VIEGAS et al. 1997, 2000; NOVATO, 2000; BOMBARDELLI et aL, 2005).

The above context is corroborated when we look at the definition of the term "carcass or clean body" in different works, for example the study by Pouey and Stingeling (1996) defines this term as whole gutted fish, while Santos *et al.* (1995) refers to the carcass as a whole fish and Contreras-Guzmán (1994) refers to the torso without the head, viscera and fins, but with skin.

3.3 Chemical composition and nutritional aspects of fish

Fish is the most recommended food of animal origin in the human diet and can be recommended for people of any age, especially children, adolescents and the elderly, as well as convalescent patients, due to its chemical composition, texture and sensory quality (GERMANO et al, 1998; ZAITSEV et aL, 2004; ORDÓNEZ et aL, 2005 a; DAMASCENO, 2009).

Fish meat has a high biological protein value with high digestibility. It also has an excellent quality lipid fraction with a high content of omega-3 polyunsaturated fatty acids, especially eicosapentaenoic acid (EPA) and docosahexaenoic acid (DHA), a low level of cholesterol and high levels of fat-soluble vitamins (A, D, E and K) and the

B complex in its muscle (SANCHEZ, 1989; LEDERER, 1991; GERMANO et al, 1998; OGAWA; MAIA, 1999; FENNEMA, 2000; OETTERER , 2002; KUBOTA; EMANUELLI, 2004; XAVIER, 2009).

This food is also an excellent source of minerals such as calcium (tricalcium phosphate, calcium carbonate), phosphorus, iron and iodine (marine fish) and active catalysts represented by metals (Cu, Mn, Zn, Co, Cr, Mo and V) and metalloids (F, I, Se) used in the metabolism and to maintain the health of mammals and humans (OGAWA; MAIA, 1999);

FENNEMA, 2000; OETTERER , 2002; KUBOTA; EMANUELLI, 2004 RODRIGUES et al. 2004)

Knowledge of the chemical composition of fish is important in assessing nutritional quality, with the aim of providing information for dietary decisions by identifying sources of essential nutrients, foods that reduce the risk of chronic diseases, nutrients that serve as a basis for the conception of functional food (RODRIGUES et al., 2004; MENEGASSI, 2011).

In addition, determining the chemical composition provides important data for processing the raw material, in terms of product standardisation, as this information is used during the classification of species for the preparation of products, the monitoring of industrial processes and research into changes in chemical components, the choice of equipment and processes that can be used in processing, helping with the rational use of fish, the method of preserving fish, protection measures that increase shelf life, (CONTRERAS-GUZMÁN, 1994; SHEARER, 1994; ALMEIDA, 1998; FERREIRA et al, 2002; RODRIGUES et al, 2004; GONÇALVES, 2011). As well as improving the nutritional potential of the products, with the aim of bringing about a change in fish consumption in Brazil (MUSTAFA, 1985).

This chemical composition is extremely influenced by intrinsic factors such as the shape and size of the organism, spawning conditions, sex, age, migration, gonadal development, the part of the body to be analysed and genetic factors, as well as exogenous factors such as climate, season, abundance and type, quantity and quality of food (BUCKLEY; GROVES, 1978; FREITAS;GURGEL, 1982;

CONTRERAS- GUSMÁN, 1994; OGAWA; MAIA, 1999; MACEDO-VIEGAS et al. 2000, KUBOTA; EMANUELLI, 2004; BOCHI et al, 2008).

The relationship between knowledge of body chemical composition and the factors that influence it enables fish health to be assessed, the efficiency of nutrient transfer from food to fish to be determined and the possibility of predicting changes in carcass and fillet composition (SHEARER, 1994; ALMEIDA, 1998; FERREIRA et al, 2002).

According to Suzuki (1981), Rodrigues et al. (2004) and Santos (2011), the chemical constituents of fish include water, proteins and lipids, due to the influence of these components on the quality of the raw material for processing purposes.

3.3.1 Proteins

Proteins are macromolecules made up of the union of various amino acids, found in abundance in living beings and perform important functions such as maintaining the distribution of water between the interstitial compartments of the body's vascular system, facilitating the occurrence of chemical reactions (enzymes), transporting oxygen (haemoglobin), transmitting information (hormones), the structural composition of cells (membranes, tubules, etc.), organic defence (antibodies), control of reproduction and tissue nutrition (MAHAN; MYERS, 1995; LOPES, 1999).), organic defence (antibodies), reproductive control, as well as participating in homeostasis, blood coagulation and tissue nutrition (MAHAN; MYERS, 1995; LOPES, 1999; ALVES; WAITZBERG, 2009).

Machado (1984), Lederle (1991), Contreras-Gusmán (1994), Ogawa and Maia (1999) and Food Ingredients Brasil (2009) described that fish is a source of protein of high biological value, due to the quality and balance of essential amino acids present (histidine, isoleucine, leucine, lysine, methionine, phenylalanine, threonine, tryptophan and valine), with high digestibility, above 95 per cent.

The proteins found in fish muscle account for between 15 and 24 per cent and are divided into three main groups: sarcoplasmic proteins (20-35% of the total), which perform biochemical functions in the cells, myobribillary proteins (65-75% of the total) of the contractile system and stromal proteins (3% of the total), mainly responsible for the integrity of the muscles (PEARSONS; YOUNG, 1989; MACKIE, 1994; SIKORSKI, 1994; KUHN; SOARES, 2002; FOOD INGREDIENTS BRAZIL, 2009; SANTOS, 2011) (Table 01).

Table 01 - Protein composition in fish muscle.

SARCOPLASMIC PROTEINS	**20 - 35% in total**
▪ Glycolytic enzymes	
▪ Oxidative/ Mitochondrial Enzymes	
▪ Lysosomal enzymes	
▪ Myoglobin	
▪ Haemoglobin	
MYOFIBRILLAR PROTEINS	**60-77% in total**
▪ Muscle contraction proteins: myosin, actin, actomyosin	
▪ Regulatory proteins: tropomyosin, troponima, actinin, M protein, etc.	
▪ Support protein: connectin, protein Z, etc.	
STROMATIC **PROTEINS**	**3 % of the total**
Collagen	
▪ Elastin	

Source: Adapted from SAMS, 2001, and SANTOS, 2011

Sarcoplasmic proteins are located within the cell's sarcoplasm and are soluble in water and low ionic strength solvents (I=o.05~0.15).These proteins comprise more than 100 different types, the main ones being enzymes that are involved in the cell's metabolism, such as: a) glycolytic enzymes - they account for 70% of water-soluble proteins; b) Parvalbumin - this binds to Ca^{2+} at a ratio of 2 mol of Ca/Mol of protein; c) Myoglobin - a protein pigment that binds reversibly to oxygen and d) oxidative enzymes. Proteins that are

including in this category, have precipitation during cooking, a peculiar characteristic to be considered during processing due to its significant influence on the texture of the fish (OLIVO, 2002; MINOZZO.2005; FOOD INGREDIENTS BRASIL, 2009).

As far as stromal proteins are concerned, they are found in connective tissues and have aspects of insolubility in acid, water and solubility in neutral salt of 0.01-0.1 M concentration, whose proteins in this category are glycosaminoglycans, lipids, glycoproteins, as well as collagens and elastin, These tissues have a higher collagen content (±3%) which has a major influence on the quality of processed products, as it can cause undesirable characteristics, such as dough instability, water release and loss of the desired texture (SUZUKI, 1981; MINOZZO, 2005; SANTOS, 2011).

Myofibrillar proteins are proteins contained in the muscle cells that make up skeletal tissues and are mainly responsible for muscle contraction. They are insoluble in water and soluble in salt. Considered the main proteins in fish meat from a dietary point of view, they are responsible for maintaining the functional properties of the muscle (e.g. retention capacity, emulsifying properties, gelation, etc.), as well as containing important amounts of essential amino acids (KUHN; SOARES, 2002; OLIVO, 2002; RUSTAD, 2010).

According to Lehninger et al. (2006), essential amino acids are defined as amino acids that are not synthesised by organisms and need to be ingested during feeding. These amino acids can be found in fish in general, making this type of food an important source of these nutrients.

It should be noted that when fish is compared with other foods such as meat, milk and eggs, it has a high content of isoleucine (10.85%), valine (11.19%), arginine (12.20%), leucine (15.93%) and lysine (18.98%), totalling approximately 69.15% of the total amino acids (Table 02) (LEDERLE, 1991; OGAWA and MAIA, 1999; ALVES and WAITZBERG, 2009).

Table 02 - Composition and percentage of amino acids in eggs, milk, meat and fish.

AMINO ACID	EGGS (mg/g)	%*	MILK (mg/g)	%*	MEAT (mg/g)	%*	FISH (mg/g)	%
Arginine	400	11,83	230	7,03	410	13,40	360,00	12,20
Cystine	130	3,85	50	1,53	80	2,61	70,00	2,37
Histidine	160	4,73	170	5,20	200	6,54	130,00	4,41
Isoleucine	360	10,65	390	11,93	320	10,46	320,00	10,85
Leucine	560	16,57	620	18,96	490	16,01	470,00	15,93
Lysine	420	12,43	490	14,98	510	16,67	560,00	18,98
Methionine	190	5,62	150	4,59	150	4,90	180,00	6,10
Treonima	330	9,76	290	8,87	280	9,15	280,00	9,49
T riptophan	110	3,25	90	2,75	80	2,61	60,00	2,03
Tyrosine	270	7,99	350	10,70	210	6,86	190,00	6,44
Valine	450	13,31	440	13,46	330	10,78	330,00	11,19

* Percentages of the amino acids contained in the food Source: Adapted from GUHA (1962)

3.3. 2Lipids

Lipids are organic compounds made up of groups of fatty acids, long-chain carboxylic acids and unbranched chains with single or double bonds. These compounds are insoluble in water; however, they are soluble in solvents with low polarity, such as chloroform and ether (FOOD INGREDIENTS BRASIL, 2009; TORRINHAS et al, 2009).

The main functions of lipids are to provide energy with a high calorific density (09 Kcal/g), supply essential fatty acids, transport and absorption of vitamins, mechanical protection, maintenance of body temperature, synthesis of cellular structures such as membranes, etc.

phospholipids, hormone synthesis, intra- and extracellular mediators of the immune response, participation in the inflammatory process and oxidative stress, enzymatic co-factors, electron carriers, pigments, emulsifying agents (FOOD INGREDIENTS BRAZIL, 2009; TORRINHAS et al, 2009; MEDEIROS, 2012).

Albert et al. (1998), Fennema (2000), Oetterer (2002), Suárez-Mahecha et al. (2002) and Fatch et al. (2010) stated that fish is a food with an excellent quality lipid fraction, due to the low amount of saturated fats and high amount of polyunsaturated fatty acids, mainly a-linolenic acid ((JÚ-3), the precursor of eicosapentaenoic acid (EPA) and docosahexaenoic acid (DHA), and linoleic acid (co-6), the precursor of arachidonic acid, which is involved in cholesterol control, a characteristic not found in any other food of plant or animal origin.

The lipid content found in fish muscle represents around 0.1 to 22%, and this percentage varies according to sex, size, age, degree of maturity, reproductive cycle, water temperature, depth, habitat, season, collection area, eating habits, nutritional status, diet composition, as well as differences between species and/or between the same species (SUZUKI, 1981; JUSTI et al., 2003; RODRIGUES et al.,2004; SANTOS, 2011, MEDEIROS, 2012). This constituent, according to Medeiros (2012), comes in the form of deposit lipids, composed mainly of triglycerides (TG), and tissue lipids, consisting of phospholipids, glycolipids and sterols.

According to Jacoquot (1961), Machado (1984), Lederle (1991), Ogawa and

Maia (1999) and Lima (2002), based on the lipid content present in the muscle, fish can be classified as lean, semi-fatty or fatty fish, as shown in Table 03.

Table 03 - Classification of fish according to their lipid content.

CLASS	LIPID CONTENT (%)	EXAMPLE OF SPECIES
Lean fish	**< 2, 5%**	**Cod** *(Gadus morrhua);* **Merluza** *(Merlucius hubbsi)',* ■ *Dogfish (Carcharhinus* **sp.);** **Corvina** *(Micropogonias furnieri);* **Hake** *(Cynoscion* **spp.).**
Semi-fatty fish	**2-10%**	**Salmon** *(Salmo salar);* ■ **Mackerel** *(Scomberomorus cavalla);* ■ **Curimbatá** *(Prochilodus* **sppj;** **Lambari** *(Astyanax taeniatus);* ■ **Common Pacú** *(Mtynnis hypsauchen).*
Oily fish	**>10%**	■ **Tuna** *(Thunnus thynnus)-,* **Sardines** *(Sardinella brasiliensis);* **Mullet (Mugil sp.);** ■ **Mandi** *(Pimelodus ciarias)* **Mapará** *(Hypophthalmus dentatus)*

It should be noted that the consumption of polyunsaturated fatty acids, such as palmitoleic, linoleic, linolenic and arachidonic acids, during the diet has shown numerous advantages for human health, proven by various studies, as a tool for preventing cardiovascular diseases, coronary atheroma, especially in people with depressive and mental disorders (CONNOR; CONNOR,1997; ALBERT et al.,1998; SU et al., 2003).

The trigger for the advantages of fatty acids is based on the most active biotransformation capacity (long and unsaturated) within organisms during the functions of maintaining the homoeostatic balance and structuring cell membranes, brain and nerve tissues (FERRETTI et aL, 1994; OSSA 1995).

3.3.3 Humidity (Water)

Water is the most abundant constituent of fish, accounting for around 60 to 85% of the total chemical composition, and this percentage varies depending on the species, time of year, age, sex, nutritional status and parts of the fish to be analysed (MACHADO, 1984; OLIVO, 2002; MEDEIROS, 2012). This constituent, according to Medeiros (2012), comes in the following forms:

■ *Uvre* water refers to water acting as a dissolving-solvent medium in the transport of nutritional components and metabolised products, participating in the maintenance of electrolyte balance and controlling osmotic pressure in muscle and connective tissue structures. This nature of the solvent usually favours its freezing at a temperature of -1 to -2°C.

■ *Constitutive water: this* refers to the water that binds proteins and carbohydrates. This water is not a solvent, which is why it is difficult to freeze at the same temperature as free water.

It is extremely important to point out that the moisture content found in fish is inversely proportional to the amount of fat, i.e. lean fish (< 2.5% fat) have a higher amount of water, while fatty fish have a lower moisture content (> 10%) (Table 04) (JACOQUOT, 1961; SIKORSKI etaL, 1994; OGAWAe MAIA, 1999; MINOZZO, 2005; MEDEIROS, 2012).

Table 04 - Amount of moisture, protein and lipids in some fish species.

FISHING	SCIENTIFIC NAME	MOISTURE (g/100g)	PROTEIN (g/100g)	LIPIDS (g/100g)
Lean fish				
Jaraqui	*Semaprochilodus* sp.	76,55	20,16	1,43
Corvina	*Micropogonias fumieri*	79,40	18,60	1,60
Hake	*Cynoscion striatus*	80,60	15,50	1,10
Acari-bodo	*Pterygoplichthy multiradiatus*	82,36	16,20	0,29
Semi-fatty fish				
Curimbatá	*Prochilodus* spp	75,60	17,40	6,00
Lambari	*Astyanax taeniatus*	72,20	15,70	9,40
White hake	*Cynoscion* spp.	79,60	16,30	4,60
Common Pacú	*Mtynnis hypsauchen*	74,56	18,41	2,97
Oily fish				
Mandi	*Pimelodus ciarias*	65,15	17,22	15,51
Mapará	*Hypophthalmus dentatus*	64,91	11,37	20,94
Matrinchã	*Brycon* sp	60,00	19,30	18,70
Sardines	*Sardinella brasiliensis*	58,50	18,50	20,70

Source: GONÇALVES E MENEGASSI (2011) and NEPA/ UNICAMP (2011)

Fogaça et al. (2009) emphasised that information on the water content present in the muscle of a particular species is very important data for processing and preserving it, as this water content directly influences the stability, quality and composition of the final product.

3.4 Functional properties: water retention capacity and water loss in fish

Functional properties are defined as all the physico-chemical and non-nutritional properties that characterise food and influence its use during the industrialisation process, in determining final quality, as well as decisively in the economic aspects of final products. These properties are influenced by the percentages, types and behaviour of certain chemical components (moisture, lipids and proteins) (OLIVO, 2002; ORDÓNEZ et al, 2005 b).

According to Olivo (2002) and Roça (2006), the fundamental functional properties for processing and preserving food of animal origin such as fish are: water retention capacity, emulsification and gelation, colour, flavour, cohesion, structure and texture.

Among these properties, the water retention capacity (WRC) is directly related to the softness of the product, its appearance, palatability and juiciness, as well as the quality of fish destined for direct consumption and industrialisation (BRESSAN, 1998; MENDES et al., 2003; ROÇA, 2006; SOUZA, 2006; CASTRO, 2007).

Water retention capacity can be defined as the ability of meat to retain its moisture or water during the application of external forces, such as cutting, heating, grinding and pressing and/or centrifugation of muscle tissue (FLORES;BERMELL, 1984; SÁ, 2004; ROÇA, 2006; PINHEIRO et al., 2009), i.e. the ability of the muscle to resist water loss (LAKSHMANAN et al.,2007).

However, during a gentle application of any of these treatments, there is a certain amount of water loss, because some of the water present in the fish is in free form and bound in the meat, and around 90 per cent of the bound water is diffused in the tissue by capillarity, mainly in the intracellular spaces (OFFER; TRINICK, 1983; ROÇA, 2006).

The RAC is considered an important parameter used to determine the quality of meat from animal feed, since the quality of the product obtained is significantly affected by the water content and its distribution in the muscle (OLSSON et al, 2003).

However, the CRA, according to Honikel and Hamm (1994), is not characterised as an objective parameter due to the lack of a real value for this property, so it only expresses a tendency. This is because it is a dynamic process and is easily influenced by changes in protein configuration, fluid distribution in intra and extracellular spaces, pH, ionic strength and physical forces (pressure and heat) during the freezing, thawing (*drip loss)* and *cooking (cooking* /oss) process (JONSSON et al, 2001, CASTRO, 2007), as well as physical conditions of the fish, state of sexual maturation, development and other factors, but these last three factors have little influence on the CRA (HUSS, 1988).

When tissues have a low CRA, there is a great loss of moisture and weight during storage. This happens when the muscle is in the process of *rigor mortis. As* storage time passes, its CRA drops, directly influencing important meat quality parameters such as texture, appearance and preservation stability, specifically cell distribution (HUSS, 1988; LAKSHMANAN et al, 2005; VENTURINI et al, 2007).

It is important to emphasise that water retention capacity is correlated with certain physical and chemical properties, such as pH. This situation is corroborated by Ofstad et al, (1996a) who reported the influence of pH on WRC in cod and salmon and ensured that low pH is not the only reason for the reduction in WRC.

CHAPTER 4

DESCRIPTION OF THE SPECIES UNDER STUDY

4.1 Taxonomic classification

- **Kingdom:** Animalia
- **Subkingdom:** Metazoa
- **Phylum:** Chordata
- **Subphylum:** Vertebrata
- **Super class:** Pisces
- **Class:Osteichthyes**
- **Infraclass:Teleostei**
- **Superorder:** Actinopterygii
- **Order:** Siluriformes
- **Family:** Ariidae
- **Genre:** *Sciades*
- **Species:** *Sciades proops* (Valenciennes, 1840) (Figure 01)

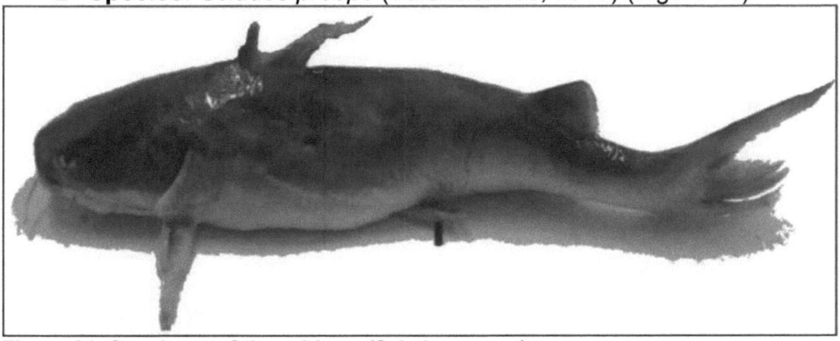

Figure 01- Specimen of the uritinga *(Sciades proops).*

4. 2Body characteristics

Elongated body with uniform grey or grey-blue colouration towards the dorsal region and white in the ventral region. It has a dorsal and pectoral region with a lumbar spine and adipose fin present, with 16-19 soft rays on the anal fin. The head is compressed, with a clearly visible, rough, exposed bone shield. It has 14-19 total gill tracks on the first gill arch and a snout with a slightly rounded deep mouth. It has a mandible and two mentalis, the pectoral ex-fin beads with villiform teeth, palate distributed a U-shaped patch (ALCALÁ; SÁNCHEZ-DUARTE, 2011).

4.3Size and weight

According to Cervigón (1991) and Novoa (2000), S. *proops* can reach a length of over 01 m and a maximum weight of 15 kg. However, it generally has a total length of between 60-80 cm.

4. 4Food and eating habits

It is a carnivorous species, whose diet is made up of various species of fish, molluscs and crustaceans, especially crabs. There is no food preference due to the time of year or length classes (GUEDES; VASCONCELOS FILHO, 1980; NOVOA, 2000).

4.5Habitat and abundance

Species belonging to the Ariidae family occur in tropical and subtropical coastal areas in estuarine waters, coastal lagoons and coastal marine environments up to 20 metres deep, mainly on muddy seabeds, and are abundant in coastal waters with muddy bottoms and shallow depths (ARAÚJO, 1988; CERVIGÓN, 1991; SANTO et aL, 2005).

4. 6Occurrence

This is a species of catfish that has a wide distribution, occurring along the entire north-east coast of South America and the Caribbean Islands in rivers that drain into the Atlantic and in estuaries from Colombia to Brazil, in the state of Maranhão (TAYLOR; MENEZES, 1978; REIS et aL, 2003).

4. 7Reproduction

These fish reproduce sexually without parental care and through external fertilisation, with a ratio of 3 females for each male. Their reproductive period occurs between the months of October and May, with a pronounced peak in activity in February and a lower peak between November and April, indicating that the females do not feed well during the reproductive season, prior to spawning (ETCHEVERS, 1978; REIS, 1986; FÁVARO et aL, 2005; CANTANHÊDE et aL, 2007; AZEVEDO; CASTRO, 2008).The spawning site is always close to the mouths of rivers and lacunar regions (FIGUEIREDO; MENEZES, 1978; BRITO, 2005). The eggs are incubated by the male, who does not feed for about two months (LE BAIL et al., 1984).

4.8 Commercialisation

The uritinga (S. *proops)* is one of the main species favoured by consumers in Maranhão and is easily found in retail outlets (markets, fairs and supermarkets). It should be noted that among the retail outlets, the fairs stand out because they offer greater access to the product, due to their location being closer to the consumer market.

The species is sold as whole fish or chilled gutted fish, with a price of between R$8.00 and R$10.00, which is influenced by the demand for the product.

CHAPTER 5

METARIAL AND METHODS

5.1 Collection of past data

A bibliographical survey was carried out, mainly on morphometrics, fillet yields and the chemical composition of marine species, using books, scientific journals, technical reports, theses, dissertations, monographs and electronic documents in the Scielo databases, the Capes Periodical Portal and the websites of institutions that are directly and indirectly linked to the subject.

5.2 Sampling

Thirty specimens of nettle *(S.proops)* were purchased from the retail trade in the city of São Luís-MA and transported in an isothermal box with flaked ice to the Food Physics and Chemistry Laboratory at the State University of Maranhão for analyses of morphometric characteristics, meat yields and physicochemical and bromatological parameters.

5.3 Analysis methods

5.3.1 Morphometric characteristics

Morphometric characteristics were measured, such as: total length (TL)[1] , standard length (SL)[2] , head length (HS)[3] , zoological length (ZL)[4] and dorsal height (DL), as shown in figure 02, using an ichthyometer and a pachymeter as an auxiliary tool.

---

[1] Measure from the cranial end to the end of the caudal fin
[2] Measure from the cranial end to the caudal peduncle
[3] Measure from the **cranial end to the back of the operculum**
[4] Measurement from the facial **end to the furcal end of the caudal fins**

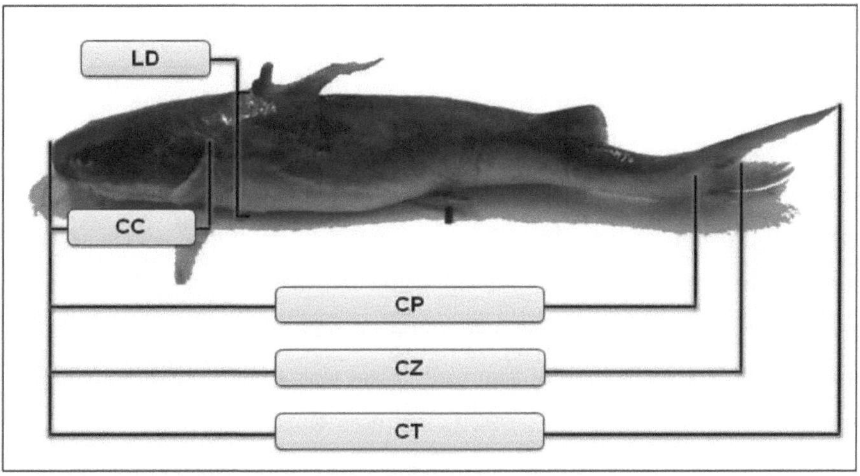

Figure 02 - Morphometric measurements taken of uritinga specimens (S. *proops)*
Legend: CP - Standard length, CZ - Zoological length, CT- Length total, CC - head length and LD - dorsal height

The morphometric measurements were used to analyse the main body ratios used for processing species with industrialisation potential: head length/standard length (CC/CP), dorsal height/standard length (LD/CP) and dorsal height/head length (LD/CC), which were expressed as proposed by Leis and Trnski (1989) and Hurlblt and Clay (1998).

5.3.2 Fillet yield

The fillets were obtained using appropriate knives and scissors as an auxiliary tool, applying a cut in a cranial-caudal direction, which resulted in fillets with skin on the right side (FCPD) and left side (FCPE), weighed, followed by the removal of the skins, These fillets were then weighed, followed by the removal of the skins from these fillets with the help of a knife, resulting in skinless fillets on the right (FSPD) and left (FSPE) sides. After these fillets were obtained, they were weighed and, at the same time, the residues[5] (PC) were weighed, with a greater focus on the weight of the head (PCAB). It should be emphasised that the filleting process was carried

[5] Refers to the fins, viscera, head and skin resulting from the filleting process

out manually by the same person.

After weighing, the waste was discarded and the fillets were packed in Styrofoam trays and kept chilled at 5°C for a maximum of one day for analyses of their physico-chemical parameters.

All these procedures were used to calculate the yields of fillet with skin on the right and left and skinless fillet on the right and left, as well as determining the percentile of carcass with head[6] and without head[7] and total residues, which were complementary to the evaluation of fillet yield. Fillet yields were obtained using the recommendations of Morais *et al.* (1992), Contreras-Guzmán (1994), Reidel et *al.* (2004) and Souza and Inhamuns (2011), which consists of applying the following formula:

$$\eta = \frac{Pf}{Pi} * 100$$

Where:

η = Yield

Pf= Fillet weight

Pi = Weight of whole fish

5.3. 3Physico-chemical parameters

5.3.3. 1Sample preparation

In order to determine the physicochemical parameters, 30 skinless fillets were used, without distinguishing the side obtained, which received three types of sample preparation procedures based on the method described by Fogaça et. al.(2009):

1)To analyse the parameters of moisture, protein, ether extract, ash and water retention capacity (WRC) under pressure - 10 samples were used, which were

[6] Term also used to refer to gutted fish

[7] Headless carcass term used for whole fish, without viscera and head

crushed and homogenised to facilitate the analysis of each parameter under study.

2) To assess *drip* loss, 10 skinless fillets were frozen at -18°C until the time of analysis.

3) To analyse water loss due to *cooking (cooking loss),* 10 frozen samples were used and thawed in a refrigerator to carry out the analysis.

5.3.3. 2Physico-chemical determinations

To assess the physico-chemical parameters, tests were carried out on moisture (UM), ash (CZS), crude protein (PB), ether extract (EE), pH, water loss due to cooking (*Cooking Loss),* drip loss *(Dríp loss) and* pressure.

The following methods were used to determine each parameter:

■ **Moisture (UM):** the gravimetric method described by the Association of Official Anlytical Chemists - A.O.A.C. (2000) was applied. This method consists of placing the crushed and homogenised sample in Petri dishes in an oven at 105°C to completely remove the moisture contained in the dish, then transferring it to a desiccator to cool the dishes and weighing it again with the dry sample.

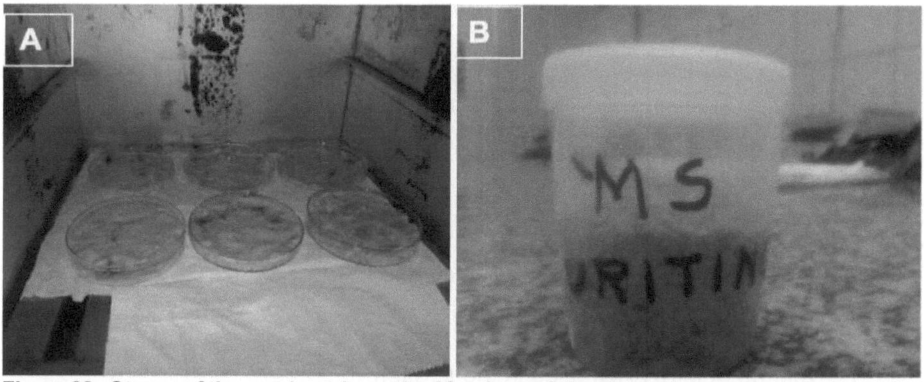

Figure 03 - Stages of the gravimetric method for determining the moisture content of samples. A) Petri dishes with wet samples in the oven at 105°C and B) Dry sample obtained.

The moisture content of the samples was expressed as a percentage obtained by applying the following formula:

$$\% \boldsymbol{Umidade} \ (\boldsymbol{UM}) = \frac{P_{AU} - P_{AS}}{PA} * 100$$

Where:

P_{AU} = weight of petri dish with wet sample

P_{AS} = weight of petri dish with dry sample

PA= weight of wet sample

■ **Crude protein (CP):** the Kjedahl method was used to determine total nitrogen, in accordance with AOAC (2000), Silva and Querioz (2004) and Fogaça et. al. (2009). This method is based on three stages: 1) Digestion - the nitrogen is transformed into ammonia and the organic compounds are converted into CO2, H2O etc.; 2) Distillation - the ammonia is separated and collected in a receiving solution; and 3) Titration - quantitative determination of the ammonia contained in the receiving solution.

The procedures carried out are based on heating the samples with sulphuric acid and a catalyst mixture at a temperature of 450°C for 5 hours for digestion, until the carbon and hydrogen are oxidised. This is followed by the addition of concentrated NaOH to the sample, which is heated to release the ammonia into a known volume of a boric acid solution, forming ammonium borate, noting that after this stage the samples have a greenish colour. The result of the crude protein content was first determined using the percentages of total nitrogen and crude protein in the dry matter. After obtaining this data, the actual percentage of crude protein was obtained using the following formulas:

■ *Determination of Total Nitrogen*

$$\% \boldsymbol{NT} = \frac{V * N * 0{,}014 * 100}{PA}$$

Where:

V = volume of 0.1 N HCl used in the titration

N = HCl Normality

PA = Sample weight in grams

0.014 = milliequivalent gram of nitrogen

- *Determination of crude protein in dry matter*

$$\%\boldsymbol{PB}_{\boldsymbol{MS}} = \%NT * 6{,}25$$

Where:

PB_{MS} = Crude protein in dry matter (%)

NT = Total Nitrogen

6.25 = nitrogen to protein conversion factor

- *Determining the real value of crude protein*

$$\%\,PB = \frac{\%PB_{MS} * MS}{100}$$

Where:

CP = Crude Protein (%)

PB $_{MS}$ = crude protein in dry matter (%) DM = amount of dry matter in the sample

- **Hydrogen Potential (pH):** A portable **pH** meter was used, using the methodology of Kirschnik (2007). This method is based on weighing 10 g of the sample, then adding 40 ml of distilled water, homogenising the mixture and filtering it through filter paper, inserting the electrode into the filtered sample, waiting for it to stabilise and then writing down the pH value determined (Figure 04).

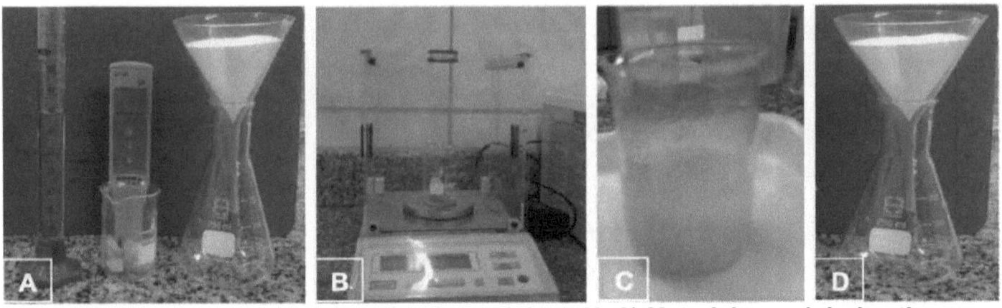

Figure 04. Kirschnik methodology for determining pH. A) Materials used during the process, B) Weighing the sample C) Homogenised sample and distilled water and D) Filtering the mixture.

■ **Water loss due to thawing *(dríp loss):*** the samples were weighed, then packed in Styrofoam trays covered with polythene film and stored in a -18°C freezer until the samples were completely frozen. The samples were then placed in a glass funnel, packed in an erlenmeyer flask and covered with cling film. They were kept at 4°C for three days, and at the end of this period the samples were weighed. The results were obtained using the following formula:

$$Drip\ loss\ \% = \frac{Pi - Pf}{Pi} * 100$$

Where:

Pi = Initial sample weight Pf = Final sample weight

■ ***Cooking* water loss*: the*** fillets were weighed, placed in plastic bags and heated to 90°C for 5 minutes, followed by cooling in cold water so that the internal temperature of the samples was between 25 and 30°C, dried on absorbent paper and weighed. The results were estimated using the following formula:

$$Cooking\ loss\ \% = \frac{Pi - Pf}{Pi} * 100$$

Where:

Pi = Weight of sample before cooking Pf = Weight of sample after cooking

■ **Water retention capacity (WRC):** approximately 1 g of the ground fillet was weighed, wrapped in filter paper, placed in test tubes and centrifuged for 10 minutes at 3000 rpm. After centrifugation, the sample was removed from the paper and weighed. The results for this parameter were obtained using the following formula:

$$CRA_{Centrifugação}\,\% = \frac{UM(\%) * Pf}{Pi}$$

Where:

UM= Moisture content of the sample

Pi = Initial sample weight Pf = Final sample weight

5.4 Experimental design

For the fillet yield analysis, a completely randomised experimental design with 30 replicates was used, with one whole fish considered as the experimental unit. The physico-chemical parameters were analysed using a 10 replicate design for each parameter.

5.5 Statistical treatment

The data obtained from the analytical procedures was organised in Microsoft Office Excel 2007 and statistically analysed in Bio Estatst 5.3. The data was subjected to the descriptive statistics parameters of measures of relative location (maximum and minimum), central tendency or position (mean) and dispersion (standard deviation and coefficient of variation), which were expressed in descriptive forms, tables and graphs.

CHAPTER 6

RESULTS AND DISCUSSION

6.1 Body characteristics

The uritinga specimens from the retail trade in the city of São Luís, MA, had the following morphometric characteristics: total length between 35.00 and 46.00 cm, with an average of 41.14 cm (±2.38); standard length between 27.40 and 38.50 cm, with an average of 33.64 cm (±2.65); zoological length, with a range between 31.50 and 40.00 cm, with an average of around 35.79 cm (±2.21). Head length ranged from 6.84 to 9.80 cm, with an average of 8.52 cm (± 0.77), and dorsal height ranged from 4.50 to 6.44 with an average of 5.40 cm (±0.48) (Table 05 and Appendices A and B).

Table 05 - Minimum, maximum and average values of the morphometric characteristics of uritinga specimens from the retail trade in the city of São Luís-MA.

VARIABLES	MINIMUM	MAXIMUM	AVERAGE*
CT	35,00	46,00	41,14 (±2,38);
CP	27,40	38,50	33,64 (±2,65);
CZ	31,50	40,00	35,79 (±2,21)
CC	6,84	9,80	8,52 (± 0,77),
LD	4,50	6,44	5,40 (±0,48)

Mean ± Deviation (n=30). CT = total length, CP = standard length, CZ = zoological length, CC = head length and LD = dorsal height.

According to Reist (1985), Huang and Liao (1990) and Santos (1994), these characteristics are considered important parameters in the process of selecting species to be used for processing from an economic and production point of view. Based on this information, it is possible to determine the minimum sizes of species that can be processed and, consequently, to provide subsidies for production and economic and financial projections for fish processing industries (VALENTI et al., 2000; MACHADO; FORESTI, 2007).

6.1. 1Morphometric relationship : head length and standard length

According to the morphometric relationship between head length (HS) and standard length (SL) (Figure 05), it was found that the specimens under study had an average HS of 25.25% (± 1.87), with a range from 22.53 to 29.76% of the SL, which classified them as moderate-headed species, according to Leis and Trinski (1989), with greater utilisation of the trunk.

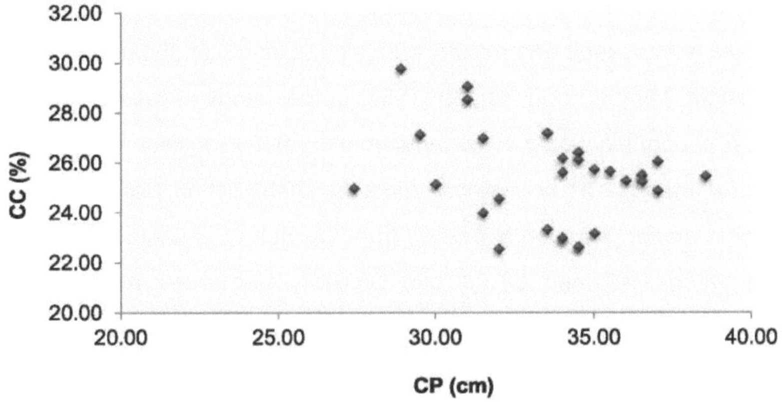

Figure 05 - Proportion of head length (HS) in relation to standard length (SL) of *Sciades proops*

The results found for the CC/CP ratios of the species under study were similar to the values described by Szpilman (2000) when he reported on the general characteristics of the curema mullet *(Mugil curema)* and mullet *(Mugil liza),* which found head lengths of around 27 to 28% and 24 to 26% of CP, respectively; Vilas Boas (2001) when assessing the morphometric CC/CP of matrinxã *(Brycon cephalus)* in different weight classes, found an average value of 24.00%, and Reidel et al (2004) when characterising the body indices of curimatá *(Prochilodus lineatus) and* piavuçu *(Leporinus macrocephalus)* observed ranges of 22.00% to 24.00% and 25.00 to 26.00%, respectively.

The values obtained for the CC/CP of the uritinga were higher than those found for the piracanjuba *(Brycon orbignyanus),* according to Freato et al. (2007), which showed CC at around 20.70 per cent of CP, *Trichomycteris jacupiranja,* as described by Wosiacki and Oyakawa (2005), which provided a range of 20.70 to 22.40 per cent for this ratio, and some fish species from the Exclusive Economic Zone in the south-east-south region of Brazil, such as *Thyrsitops lepidopoides* (SC between 3.40 and 3.80 per cent of CP) and *Benthodesmus tenuis* (SC between 7.10 and 7.80 per cent of CP), according to Bernardes et al.(2005).

On the other hand, these values were lower than those observed by Bombardelli and Sanches (2008) in their description of the morphometry of the armado *(Pterodoras granulosas),* which expressed a head length of 35.00 to 37.00% of the standard length; Miranda and Ribeiro (1997) in their study of the processing yields of surubim *(Pseudoplatystoma coruscans)* from the São Francisco river, found a 36.00% index for the CC/CP correlation; Almeida (2010) when assessing the CC/CP body ratio in mandi *(Pimelodus britskii)* from the Salto Santiago-Rio Iguaçu reservoir, found values between 27.50 and 38.06%; Bernardes et al.(2005) characterised some of the fish species in the Exclusive Economic Zone of the south-east-south region of Brazil as *Lopholatilus villarri* (CC between 32.00 and 33.00% of CP) and *Pagrus pagrus* (CC between 31.30 and 33.30% of CP).

In this context, it can be seen that the CC/CP body ratio values obtained from the uritinga are the result of specific habitat conditions and the organism's adaptations to the environment. Taking this into account, *S.proops* showed a CC/CP ratio that was more similar to that of freshwater species than other marine species such as snapper, which may be due to the lack of information on this ratio for most species.

According to Reidel et al (2004) and Bombardelli and Sanches (2008), the head length/standard length ratio (HC/SL) is one of the most important parameters for industrialising fish, as it directly influences the percentage of fillet and waste utilisation, since the size of the fish's head is considered an inedible part, so its elevated size can cause losses in fillet yield.

6.1. 2Morphometric relationship : dorsal height and standard length

With regard to the relationship between dorsal height (DL) and standard length (SL) (Figure 06 and Appendices A and B), it can be seen that the uritinga specimens exhibited DL values ranging from 12.43 to 18.81%, the average of which was around 16.10% (±1.36) of the SL, so it is characterised as a species with an elongated body, according to the criteria established by Leis and Trnski (1989), and should have a higher trunk yield.

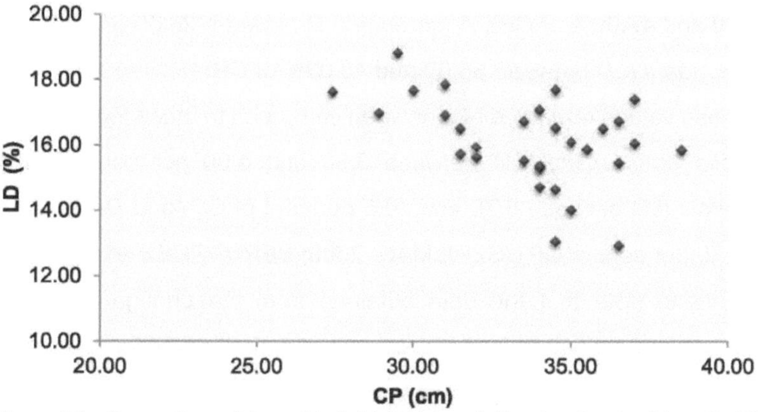

Figure 06 - Proportion of dorsal height (DL) in relation to standard length (SL) of *Sciades proops.*

The LD/CP values obtained for S. *proops* in the present study were similar to the data mentioned by Nakatani et al. (2001) and Almeida (2010) for some species of the Pimelodidae family *(Pimelodus britskii, Pimelodus maculatus and Pimelodus ortmanni). britskii, Pimelodus maculatus and Pimelodus ortmanni*), which showed dorsal height of around 16.48 to 26.29% of HC; and Bernardes et al. (2005) for the species *Malacathus plinieri,* a fish from the Exclusive Economic Zone of the south-east-south region of Brazil, which showed LD of between 13.00 and 19.00% of HC.

The results shown by the uritinga for this ratio were lower than the values observed by Szpilman (2000) in dory (*Selene setapinnis) (*LD between 43.00 and 56.00% of CP), bullseye (*Seriola dumerlli)* (LD between 23.00 and 26.00% of CP) and yellowtail flounder (LD between 48.00 and 59.00% of CP); Vilas Boas (2001) in matrinxã (*Brycon cephalus) (LD of* 34.00% of the CP); Bernardes et al.(2005) in some species of fish from the Exclusive Economic Zone in the south-east-south region of Brazil, such as *Epigonus telescopus* (LD between 21.20 and 26.30% of CP), *Lopholatilus villarri* (LD between 28.00 and 30.00% of CP); Freato et al. (2007) in piracanjuba (*B. orbignyanus) (LD* between 28.00 and 30.00% of CP). *orbignyanus)*

(LD of 31.70% of CP), and Kuwabara et al (2011) in some species of the *Astyanax* genus such as *A. fasciatus (LD* between 26.00 and 39.00% of CP), A. *altiparanae (LD between* 36.00 and 47.00% of CP), *A. paranae* (LD between 26.00 and 39.00% of CP) and *A. bockmanni (LD* between 35.00 and 43.00% of CP).

However, these values remained higher when compared to those found in true sardines *(Sardinella braisliensis)* (LD between 3.50 and 4.50 per cent of CP), *Percophis brasiliensis* (LD between 8.70 and 10.5 per cent of CP) and *Decaptacus plumierri (LD of* 5.70 per cent of CP) (SZPILMAN, 2000; BERNARDES et al., 2005).

It is important to note that the data obtained from the urintiga in LD/CP remained within the range estimated for species belonging to the Siluriformes order, which shows that it is a common characteristic of the same, according to Almeida (2010).

6.1. 3Morphometric relationship : head length and dorsal height

The length of the head (LC) in relation to the dorsal height (DL) (Figure 07 and Appendices A and B) showed that the specimens of S. *proops* expressed values between 139.29 and 197.03% cm, the average of which was around 158.07% (± 12.00), so it can be seen that the species under study has a moderately large and compressed head.

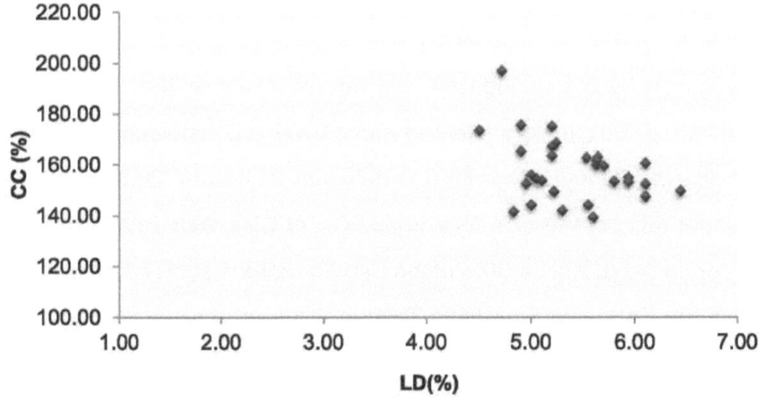

Figure 07 - Proportion of head length (HC) in relation to dorsal height (DL) of *Sciades proops*

The LD/CP values for S. *proops are* similar to the results found by Reidel et al (2004) when describing the body indices of the Curimatá *(Prochilodos lineatos),* they observed that the length of the head corresponds to around 148.00 to 154.00% of the LD, and Bombardelli and Sanches (2008) when reporting on the morphometrics of the Armado *(Pterodoras granulosos)* found that CC expressed a range of 180.00 to 193.00% of the LD.

However, the results obtained by *S. proops* remained above the values observed in the piavuçu (*Leporinos macrocephalos) (*CC 128.00 to 133.00% of the LD), piracanjuba (*Brycon orbignyanos) (CC* 93.80% of the LD), in some species of the *Astyanax* genus such as *A. clavitaeniatus (CC 95.90 to 112.10% of the LD), and* in some species of the *Astyanax* genus such as A. clavitaeniatus (CC 82.70 to 108.70% of the LD).*clavitaeniatus* (CC 95.90 to 112.10% of the LD) and *A. siapae* (CC 82.70 to 108.70% of the LD)(REIDEL et al, 2004; GARUTTI, 2003; FREATO et al., 2007)

According to Macedo-Viegas et al. (2000) and Bombardelli and Sanches (2008), the data obtained by the uritinga for CC/LD is one of the characteristic aspects of species belonging to the siluriform order. This suggests that individuals, such as the uritinga, which have high proportions of heads in relation to dorsal height may obtain a lower fillet yield when compared to the meat yields obtained by species with small, loosely compressed heads, and this species in question will consequently provide a high residue yield.

Although the CC/LD ratio is an important morphometric parameter for the industrialisation of certain species, there is still a large lack of information on the influence of this ratio on meat and carcass yield for most marine species, which makes it difficult to compare data between these species.

6.2 Fillet yield

When assessing the fillet yield of uritinga from the retail trade in the city of São Luís-MA, it was found that the specimens weighed between 318.80 and 745.30g, the average being around 541.17g (± 104.31) (Appendices C and D).

These values observed for the species under study resulted in skin-on fillets weighing between 142.30 and 376.00 g with a median value of 267.23 g (±61.01) and skinless fillets weighing between 117.40 and 320.30 g with a median of 220.90 g (±51.01) (Figure 08 and Appendices C and D). It was noted that the skinning process carried out on the fillet resulted in a reduction of between 10.51 and 23.43 % in the final weight of the skinless fillet, implying a higher percentage to be obtained in the carcass yield and total waste generated by the uritinga.

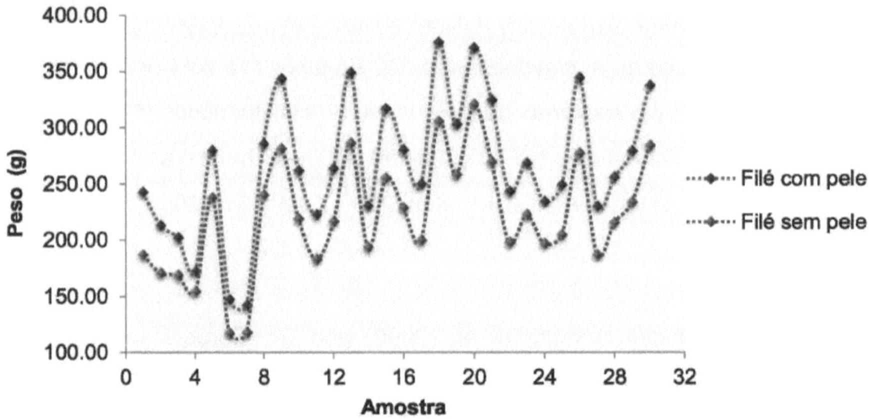

Figure 08 - Weights of skin-on and skinless fillets of uritinga specimens from the retail trade in the city of São Luís-MA.

The weights of the fillets obtained in the skin-on and skinless categories were directly proportional to the morphometric characteristics of the body (total length, head length and dorsal height).

In an intraspecific analysis of the weights of skin-on and skinless fillets considering the side on which they were obtained, it was found that skin-on and skinless fillets on the left had the highest weights, estimated at between 69.50 and 189.00g and 58.30 and 162.60g, respectively, compared to skin-on and skinless fillets on the right, which had respective values of between 69.10 and 187.00g and 55.80 and 157.70g (Figure 09 and Appendices C and D).

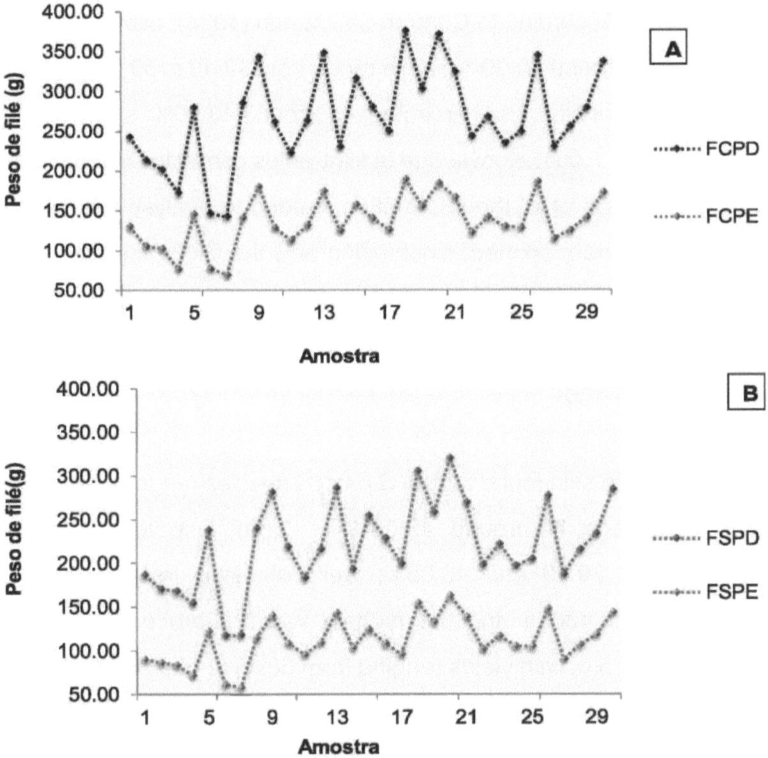

Figure 09 - Fillet weight of nettle specimens. A) Weight of fillet with skin on the right (FCPD) and left (FCPE) sides. B) Weight of fillet without skin on the right (FSPD) and left (FSPE) sides.

According to Macedo-Viegas and Souza (2004), the filleting method used in the study requires a highly skilled and experienced operator to obtain high yields.

The parameter in question, fillet yield, is one of the most commonly used parameters when assessing the methods to be used during processing to obtain the most commercial parts, as well as analysing the industrialisation potential of a given species.

Fillet yield is based on the amount of meat, made up of dorsal-lateral musculature, obtained after the filleting process, and is expressed as a percentage. Macedo-Viegas and Souza (2004) emphasised that the calculation of fillet yield is based on an intraspecific relationship with the weight of the whole fish.

In various studies, fillet yields are generally broken down into two forms: skin-on and skinless fillet yields. According to Contreras-Guzman (1994), average values for fillet yields with skin are around 50.00%, with a range from 32.40 to 59.80%. When referring to fillet yields without skin, the percentage is around 43.00%, showing that the skin reduces this yield by 7.00%.Knowledge of fillet yields correlated with carcass yields and total waste consists of all the information needed to analyse the yield of all the edible parts that can be marketed, since using only the fillet generates large quantities of waste that could be better used. It should be emphasised that carcass and total waste yield calculations are based on eviscerated body weight (CONTRERAS-GUZMAN, 1994).

6.2.1 Skin-on fillet yield

The yields of fillets with skin found for the *S.proops* species ranged from 39.89 to 53.49%, with an average of around 49.05% (± 3.89). The lowest yields, corresponding to between 39.89 and 46.05%, were obtained from specimens weighing between 300 and 430 g and the highest values were obtained from specimens weighing over 625 g, with yields ranging from 50.45 to 53.49% (Figure 10 and Appendix Ce D).

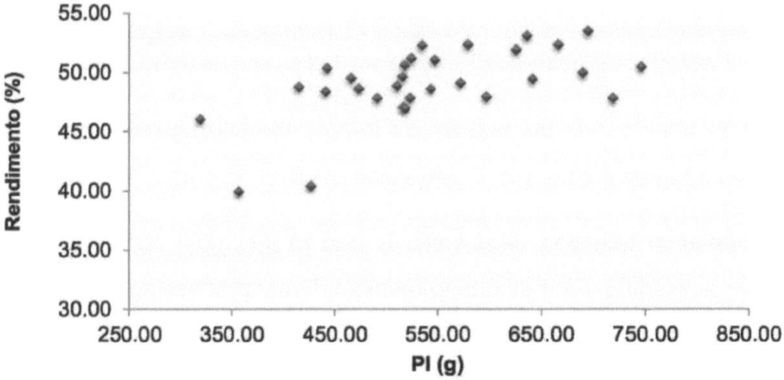

Figure 10- Ratio of skin-on fillet yield to whole fish weight (WF) for the species under study.

Based on the information provided by Contreras-Guzman (1994), the data obtained for uritinga corresponded to the expected values for marine species, which are between 32.80% and 59.80%.When comparing the yields of fillets with skin obtained with values from other species, it can be seen that they were similar to the data described by Conteras-Guzman (1994), Bruschi (2001) and Souza and Inhamuns (2011) for the following species: maria mole (*Cynoscion striatus)*, goete *(Cynoscion gatucupa)*, tuna *(Katsuwonus pelamis)*, mackerel (*Scomber scombrus)*, anchovy (*Pomatomus saltatrix)*, catfish *(Trachysurus barbus)*, shiner *(Trachurus trachurus)*, mullet *(Mugil brasiliensis)*, snapper *(Lutyanus purpureus)*, sardine (*Triportheus auritus)*, mapará *(Hypophythalmus edentatus)* and jaraqui *(Semaprochilodus insignis) (Table* 06).

Table 06 - Skin-on fillet yields of some marine and river species.

CRITERION ANALYSED	VULGAR NAME	SCIENTIFIC NAME	FILLET YIELD (%)	SOURCE
Fillet yield higher than that found in uritinga (η > 53.49%)	Albacore	*Germo alalunga*	56,80	You will contain-Guzman (1994)
	Curimbata	*Prochilodus scrofa*	59,00	
	Manjuba	*Anchovella hubsi*	57,50	
	Traíra	*Hoplias malabarucus*	58,60	
	Real sardines	*Sardinella brasiliensis*	54,30	
	Hake	*Macrondon ancylodon*	54,20	
Fillet yield similar to that found in uritinga (η between 39.89 and 53.49%)	Soft Mary	*Cynoscion striatus*	41,90	Bruschi (2001)
	Goete	*Cynoscion gatucupa*	44,30	
	Tuna	*Katsuwonus pelamis*	51,10	
	Horsetail	*Scomber scombrus*	43,50	You will contain-Guzman (1994)
	Anchovy	*Pomatomus saltatrix*	50,00	
	Catfish	*Trachysurus barbus*	49,90	
	Pee	*Trachurus trachurus*	49,00	
	Mullet	*Mugil brasiliensis*	48,40	
	Pargo	*Lutyanus purpureus*	45,50	
	Sardines	*Triportheus auritus*	38,64 - 42,21	Souza and Inhamuns (2011)
	Mapará	*Hypophythalmus edentatus*	48,29-51,89	
	Jaraqui	*Semaprochilodus insignis*	39,81 -41,86	
Fillet yield lower than that found in uritinga (η <39.89%)	Hogfish	*Balistes capriscus*	37,77	Bruschi (2001)
	Palometa	*Serrasalmus spilopleura*	31,20-36,30	Santos et al (2006)
	Butterfish	*Mylossoma duriventre*	38,96 - 39,55	Souza and Inhamuns (2011)
	Piramutaba	*Brachyplatystoma vaillantii*	29,54 - 29,98	
	Surubim	*Pseudoplatystoma fasciatum*	32,37 - 32,60	
	Tambaqui	*Colossoma macropomum*	32,00 - 35,96	
	Tucunaré	*Cichla monoculus*	35,77 - 37,96	
	Sole	*Paralichthys orbignianus*	39,80	

The yields of skin-on fillets obtained from urintiga were higher than those found in hogfish *(Balistes capriscus)*, butter pacu (*Mylossoma duriventre)*, piramutaba

(Brachyplatystoma vaillantii), surubim *(Pseudoplatystoma fasciatum),* tambaqui (*Colossoma macropomum),* tucunaré (*Cichla monoculus),* sole (*Paralichthys orbignianus) and* palometa *(Serrasalmus spilopleura)* (Table 06) (CONTERAS-GUZMAN, 1994; BRUSCHI, 2001; SANTOS et al, 2006; SOUZA and INHAMUNS, 2011)

However, the yield of skin-on fillets from nettle remained below the values found by Conteras-Guzman (1994) for yellowfin tuna (*Germo alalunga),* curimbata *(Prochilodus scrofa),* manjuba (*Anchovella hubsi),* traíra (*Hoplias malabarucus),* sardine (*Sardinella brasiliensis)* and hake *(Macrondon ancylodon)* (Table 06).

It's important to note that yields vary mainly according to the external shape of the raw material, i.e. fish with fusiform shapes have high yields due to their cylindrical muscle mass, such as tunas, bonito, sardines, curimbatá and other fursiforms, while composite fish such as corvina, sole etc. are among the species with the lowest yields (CONTERAS-GUZMAN, 1994).

6.2.2 Skinless fillet yield

The yields noted for the skinless fillets of the species under study ranged from 33.05 to 46.15 per cent, with an average of 40.54 per cent (± 2.84). The highest yields, corresponding to between 40.23 and 46.15 per cent, were found in specimens weighing between 300 and 430 g and the lowest yields in specimens weighing over 625 g, which showed values between 33.05 and 36.12 per cent (Figure 11 and Appendix C and D).

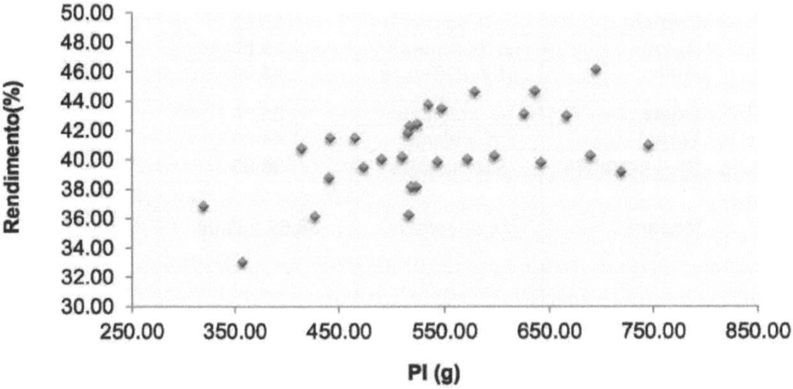

Figure 11 - Relationship between the yield of skinless fillet and the weight of the whole fish (PI) of the species under study.

When comparing the yields of skinless fillets from uritinga with values from other species, it can be seen that they were similar to catfish (T. *barbus),* snapper (L. *purpureus),* mullet *(M. brasiliensis),* goetee (C. *gatucupa),* tuna (K. *pelamis),* true sardine (S. *brasilinsis),* mapará *(H. edentatus), as shown in* Table 06 (CONTERAS-GUZMAN, 1994; BRUSCHI, 2001; SOUZA and INHAMUNS, 2011).

The yields of skinless uritinga fillets are higher than those reported by Conteras-Guzman (1994), Bruschi (2001), Santos et al. (2006) and Souza and Inhamuns (2011) for corvina (*Micropogonias furnieri),* maria mole (C. *striatus),* pig (8. *capriscus),* butter pacu *(M. duriventre),* jaraqui (S. *insignis),* matrinxã (*B. amazonicus),* sardine (T. *auritus),* piramutaba (*B. vaillantii),* surubim (*P. fasciatum),* tambaqui (C. *macropomum),* tucunaré (C. *monoculus) and* palometa *(S.pilopleura).* However, these yields of fillets with nettle skin remained below the values found for yellowfin tuna (G. *alalunga),* anchovy (P. *saltatrix),* curimbata (P. *scrofa),* hake (*M. ancylodon) and* tray (H. *malabarucus)* (Table 07) (CONTERAS-GUZMAN, 1994).

Table 07- Yields of skinless fillets of some marine and river species.

CRITERION ANALYSED	VULGAR NAME	SCIENTIFIC NAME	FILLET YIELD (%)	SOURCE
Fillet yield higher than that found in uritinga (η > 46.15%)	**Albacore**	**G.** *alalunga*	**49,00**	**Conterás-Guzman (1994)**
	Anchovy	*P. saltatrix*	**46,90**	
	Curimbata	*P. scrofa*	**50,20**	
	Hake	*M. ancylodon*	**49,10**	
	Traíra	*H. malabarucus*	**47,50**	

Fillet yield similar to that found in the (η between 33.05 and 40.54%)	**Catfish**	*T. barbus*	**38,60**	**Conterás-Guzman (1994)**
	Pargo	*L. purpureus*	**39,00**	
	Mullet	*M. brasiliensis*	**43,00**	
	Goete	*C. gatucupa*	**35,20**	**Bruschi (2001)**
	Tuna	*K. pelamis*	**44,20**	
	Real sardines	*S. brasilinsis*	**36,90**	
	Mapará	*H. edentatus*	**38,68 - 39,00**	**Souza and Inhamuns (2011)**
Fillet yield lower than that found in uritinga (η <33.05%)	**Corvina**	*M. furnieri*	**32,00**	**Conteras-Guzman (1994)**
	Soft Mary	*C. striatus*	**32,30**	**Bruschi (2001)**
	Pork	*B. capriscus*	**21,40**	
	Butterfish	*M. duriventre*	**29,88-30,65**	**Souza and Inhamuns (2011)**
	Jaraqui	*S. insignis*	**30,91 - 32,38**	
Fillet yield lower than that found in uritinga (η <33.05%)	**Matrinxã**	*B. amazonicus*	**30,50-31,66**	**Souza and Inhamuns (2011)**
	Sardines	*T. auritus*	**29,38 - 32,01**	
	Piramutaba	*B. vaillantii*	**24,04 - 24,92**	
	Surubim	*P. fasciatum*	**25,48-28,01**	
	Tambaqui	*C. macropomum*	**25,70-29,08**	
	Tucunaré	*C. monoculus*	**31,63-31,84**	
	Palometa	*S.pilopleura*	**27,00 -28,00**	**Santos et al. (2006)**

It was corroborated that removing the skin of the nettle for a more elaborate product that better meets consumer desires in terms of practicality and ease of preparation, reduced the fillet yield by 6.84 to 7.34% and based on the information observed by Souza (2004) this reduction was perfectly in line with the expected range for this item, which consists of values around 4.50 to 14.00% with an average of 7.5%, This implies an increase in the percentage of total waste generated, since the skin is considered to be waste obtained from the filleting process. Therefore, it can be seen that a significant amount of by-products can be benefited from its transformation.

6.2.3 Carcass yield

When assessing the carcass yields obtained by the uritinga, they were categorised into well-defined categories: carcass yield with and without head. The following values were found in each category (Figure 12 and Appendices E and F)

■ *Carcass yields with head:* between 88.74% for a specimen weighing 745.30g and 90.10% for a specimen weighing 665.90g, with an average of 92.17% (± 1.42).

■ *Headless carcass yield:* from 51.74% for a specimen weighing 426.20 g to 63.51% for a specimen weighing 694.10 g, with an average of 59.07% (± 2.65).

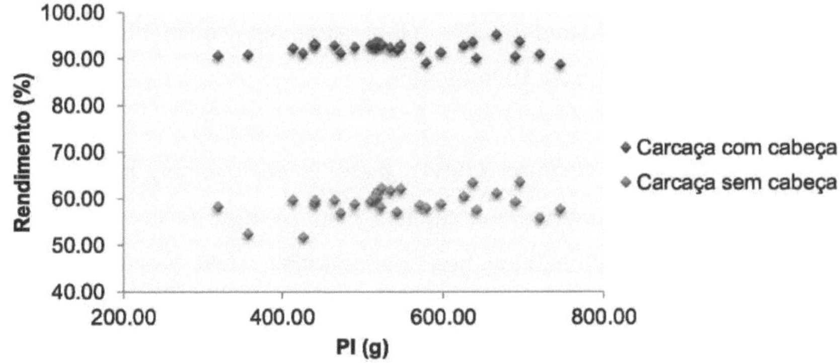

Figure 12. Ratio of carcass yield with and without head to the weight of the whole fish (PI) of the species under study.

When correlating the head and carcass yields found for S. *proops* with values found for other species, it can be seen that they were similar to curimbatá *(Prochilodus lineatus),* piavuçu *(Leporinus macrocephalus),* mapará (*H. edentatus)* and pacu manteiga (*M. duriventre)ÇVabe\a* 07) (REIDEL et aL, 2004; SOUZA and INHAMUNS, 2011). However, these remained above the values found for jaraqui (S. *insignis),* piramutaba *(B. vaillantii)* and sardine *(T. auritus), as shown in* table 08 (SOUZA and INHAMUNS, 2011).

Table 08 - Head-on carcass yields of some fish species.

CRITERION ANALYSED	VULGAR NAME	SCIENTIFIC NAME	FILLET YIELD (%)	SOURCE
Carcass yield similar to that found in uritinga (η between 88.74 and 95.10%)	**Curimbatá**	*P. lineatus*	**89,40-91,29**	**Reidel et al (2004)**
	Piavuçu	*L. macrocephalus*	**92,60 - 94,26**	
	Mapará	*H. edentatus*	**90,01 -91,56**	**Souza and Inhamuns (2011).**
	Butterfish	*M. duríventre*	**87,67 - 93,85**	
Lower carcass yield than that found in uritinga (η < 88.74%)	**Jaraqui**	*S. insignis*	**83,49 - 84,01**	**Souza and Inhamuns (2011).**
	Piramutaba	*B. vaillantii*	**84,26 - 84,99**	
	Sardines	*T. auritus*	**87,62 - 87,75**	

The headless carcass yields were lower than those reported by Souza and Inhamuns (2011) for surubim *(P. fasciatum)* (65.36 - 66.67%), matrinxã (*B. amazonicus)* (69.64 - 72.90%), piramutaba (*B. vaillantii)(6Q,Q7* - 73.2%) and sardines *(T. auritus)(67*.95 -69.68%).

Performing the decapitation process to obtain the headless carcass yield led to a reduction of between 30.00 and 37.00% in the values obtained in this regard, of which the head made up around 31.10 to 39.49%.

According to Conteras-Guzman (1994), there is an inversely proportional correlation between head percentage and carcass yield, i.e. the specimens with the highest headless carcass yields had low head percentages, while the lowest yields had high head percentages.

Excluding the head will result in an increase in the percentage of total waste generated, considering that there is a very high percentage of material of reasonably good nutritional quality that should be used to make by-products such as meal, silage, etc. (CONTERAS-GUZMAN, 1994),

6.2.4 Total waste yield

The yield of waste generated after the filleting process remained between 36.28%, found in a specimen weighing 426.20g, and 48.01%, seen in a specimen weighing 635.80g, the average of which was 40.52% (± 2.57) (Figure 13 and Appendices E and F). The percentages achieved for total residues expressed a certain convergence with that described by Conteras-Guzman (1994), which corresponds to a percentage of residues between 47.00 and 66.00%. It should be emphasised that the percentage of total waste generated depends mainly on the characteristics of the species and the skill of the operator when carrying out the filleting process.

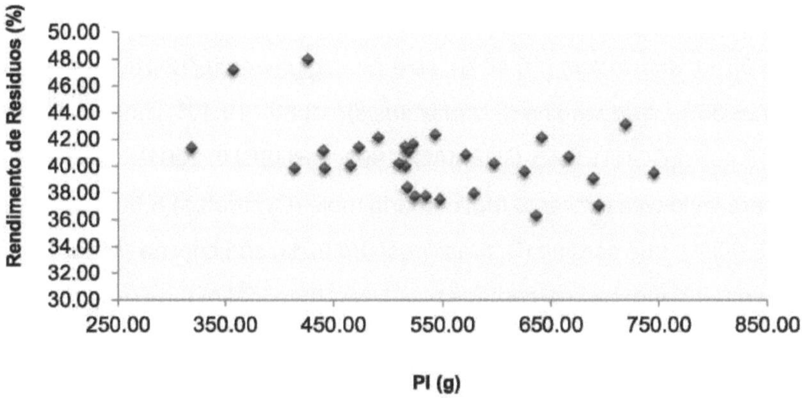

Figure 13 - Relationship between the percentage of total residues and the weight of the whole fish (PI) of the species under study.

The residue yields were categorised into two distinct classes: head and viscera, with the following values observed for each category (Figure 14 and Appendices E and F):

- *Head yield:* 30.10% for a specimen weighing 694.1 Og, to 39.49% for a specimen weighing 426.20 g. The average was around 33.10 (±2.15).
- *Visceral yield:* 6.52%, seen in a 635.80 g specimen, to 10.47%, seen in a 641.10 g specimen, in which the average remained around 8.06% (±0.98).

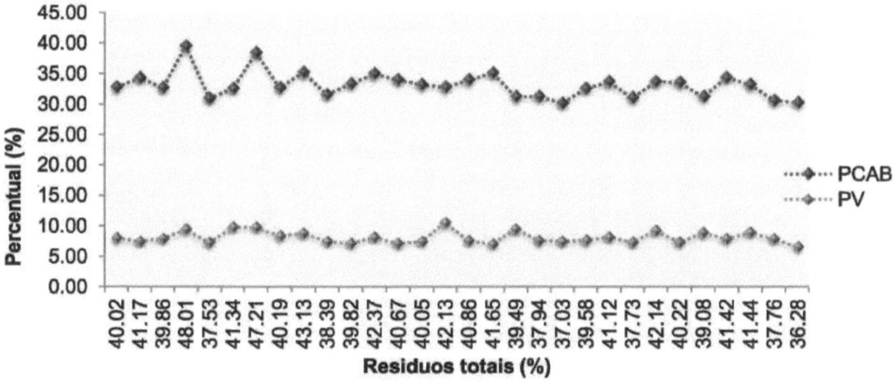

Figure 14 - Relationship between the percentage of total residues and the weight of the whole fish (PI) of the species under study.

In relation to the results observed in the head yields, it was found that this showed very high values when compared to yellowfin tuna (G. *alalunga) (*20.00%), anchovy (P. *saltatrix)* (21.50%), hake *(M. ancylodon) (12.10%), and snapper (L. purpureus) (26.35%). ancylodon)* (15.40%), sole (*P. orbignianus)* (12.10%), snapper (L. *purpureus)(26.35%),* true sardine (S. *brasiliensis)(* 13.80 - 19.40%) and mullet *(M.* bras/7/ens/s)(18.40%)(CONTERAS-GUZMAN, 1994; BRUSCHI, 2001).

The viscera yield was similar to that of maria mole (C. str/atus)(8.60%), goetee (C. *gatucupa)(8.3Q%),* true sardine (S. *brasiliensis)(8.*90%) and corvina *(M. furnieri)^* 0.00%). On the other hand, this percentage was lower than that of tuna (K. pe/am/s)(14.80%), pork (*B. capriscus){'\2,* 30%) and sea bream *(P. granulosas)^ 8.^2* - 19.03%)(BRUSCHI, 2001; BOMBARDELLI and SANCHES (2008).

Based on the information provided by Macedo-Vigas and Souza (2004), the viscera yield data corresponded to the percentage estimated for whole bony fish, which is around 11.00%.

6.3 Centesimal composition

According to the chemical composition of the fillets analysed, it was observed that the moisture content varied between 78.14 and 78.64% with an average of 78.45% (±0.14) and the crude protein content expressed values between 12.82 and 19.59%, with an average of around 17.74% (±1.93) (Figure 15 and Appendices G and H).

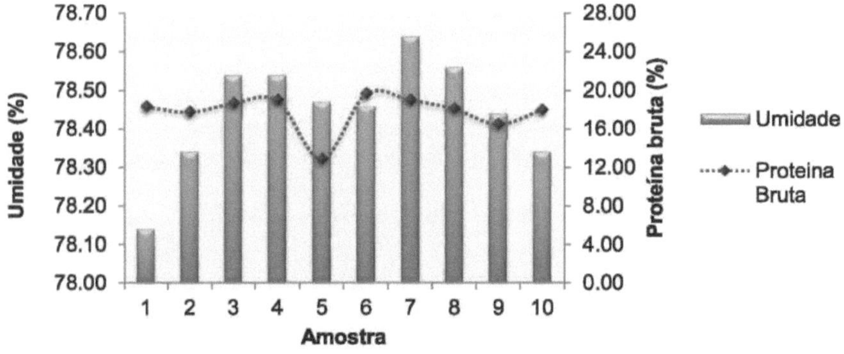

Figure 15-0 Moisture content (%) and crude protein (%) in *Sciades proops* **fillets.**

The moisture values found in the uritinga fillet are similar to those found in other fish species such as trout *(Oncorhynchus* SSP.), white hake (P/ag/osc/ons *squamossissimus),* goosefish *(Cynoscion petronus),* mackerel (*Diplectrum radiale),* hogfish (*Baliste caro)* and mullet *(Mugil cephalus),* which have the following contents, respectively: 78.20%, 78.42%, 78.50%, 78.69% and 78.40% (ORDÓNEZ et al. 2005 a; MORAES et al, 2004; NEPA/UNICAMP, 2011). This percentage is also compatible with the moisture content of other fish such as white shrimp *(Litopenaeus vannamei)* and squid *(Loligo vulgaris),* which expressed values of 78.58% and 78.51%, respectively (KIM and LALL, 2000; CEZARINI, 2005).

The average percentage of humidity recorded in this study was around 3.58% more than that recommended by the Food Technology department of the Federal University of Santa Catarina (ALMEIDA, 2010) for fish meat, which corresponds to 74.87%. According to OGAWA and MAIA (1999), this excess water in the intramuscular structure contributes to the development of microorganisms naturally present on the surface of the fish, which will speed up the deterioration process.

Crude protein showed an average value (17.74%) compatible with the protein percentages published in TACO 2011 for some species

such as whiting *(Curimata* ssp.), white hake *(Plagioscions squamossissimus),* mandi *(Pimelodus ciarias) and* roncador *(Anisotremus virginicus),* which showed respective values of 17.40 per cent, 17.42 per cent, 17.92 per cent and 17.90 per cent.
The species under study is a food source rich in protein, and this information must be taken into account for its inclusion as a functional food in the human diet, as well as the choice of the most appropriate processing, conservation and storage techniques to be used for the species, in order to maintain the qualitative and quantitative nature of this nutrient, since it is considered one of the essential nutrients that play important roles in the mobilisation and formation of body tissue.

Knowledge of the chemical composition of fish makes it possible to classify it in the major groups of food of animal origin, according to its water, lipid, protein and mineral content.This information can be used as an auxiliary tool in the standardisation of food products, to support dietary decisions, to monitor industrial processes and research for economic and technological optimisation, as well as to determine the yield for obtaining by-products such as fish protein concentrates

(CPP), fishmeal and other fish products, whether for direct or indirect consumption (JACQUOT, 1961; STANSBY, 1968; MACHADO,1984; SIKORSKI.1990; CONTRERAS-GUZMAN, 1994; HUSS, 1995)

According to Bandarra et al. (2001), a great deal of information is now available on the chemical constitution of various fish from marine, estuarine and freshwater environments. However, in some cases this data needs to be re-evaluated and updated according to new, more practical, reliable and economically trustworthy analytical methods, such as the procedures established by the Association of Official Analytical Chemists (A.O.A.C).

6.4 Physical properties

6.4.1 Hydrogen Potential (pH)

It should be noted that the hydrogen potential observed in the muscle tissue of the uritinga remained between 7.00 and 7.70, with an average of 7.40 (±0.31) (Figure 16 and Appendices G and H).

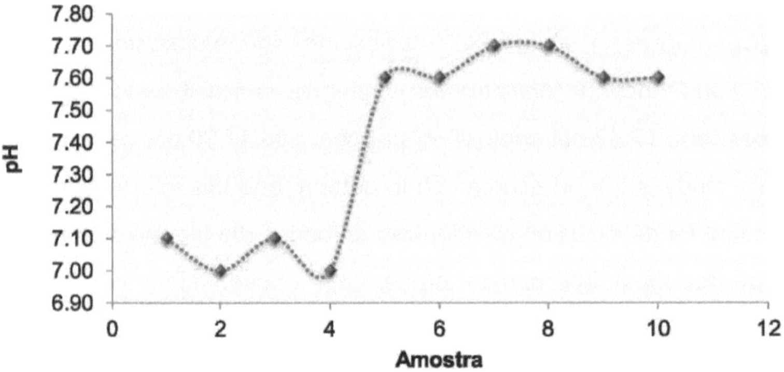

Figure 16 - Variation in hydrogen potential (pH) in the species under study

Hydrogen potential is an estimate of the concentration of acids or alkalis present in a solution (PARSONS et al, 1984). This parameter is responsible for maintaining the osmotic balance of cells, preventing denaturation of enzymes, which would compromise the integrity of the metabolism of marine organisms; determining chemical speciation in the degradation and fate of chemical elements and various compounds (CORDEIRO, 2006).

Determining pH is an important factor in assessing the quality of various foods, such as fish, which is considered to be a low-acid food, i.e. one with a pH greater than 4.5 (SAKUMA et al, 2005; TAVARES and MORENO, 2005; NOLLET and TOLDRÁ, 2010).

This concentration of hydrogen ions is normally altered when the amount of ATP produced by the anaerobic glycogen in the fish muscle is insufficient to compensate for the losses resulting from its hydrolysis by the sarcoplasmic ATPase, This results in the accumulation of lactic acid, which causes a decrease in pH, ranging from 5.6 to 5.8 in red-fleshed fish and 6.0 to 6.2 in white-fleshed fish, inhibiting various enzymes, especially phosphorylase (KIETZMANN et al, 1974; CHEFTEL and CHEFTEL,1976; OGAWA, 1999).

As the pH decreases, the proteins become denatured, the water retention capacity decreases and hydrolytic, oxidative or fermentative decomposition of the muscle is favoured, which influences the quality of the product (CONTERAS-GUZMAN, 1994; SAKUMA et al, 2005).

Based on the Regulations for the Industrial and Sanitary Inspection of Products of Animal Origin (RIISPOA) and Ordinance No. 185 of 13 May 1997 on the technical regulations for the identity and quality of fresh fish (whole and gutted), the pH results expressed for uritinga fillet indicated that the products did not fall within the range established for this parameter, considering that the pH values for *fresh* fish range from 6.5 to 6.8, it is clear that the raw material was not of good quality.

A situation similar to that found in uritinga (pH= 7.00 to 7.70) has also been seen in white hake (*Plagioscion squamosissimus)* and hake *(Macrodon ancylodon),* which had a pH of 7.07 and 7.50, respectively (SANCHEZ et al., 1990; TELLES et al., 2010).

It is important to corroborate that, according to Sikorski et al. (1990), the ph values achieved were influenced by species-specific characteristics such as the type of muscle analysed, the physiological state of the muscle, the initial glycogen content, the speed of post-mortem reactions resulting in the formation of lactic acid and subsequently its oxidation and disappearance.

6.4. 2Water loss

When analysing the percentage of water loss in uritinga fillets, the influence of two very different processes was observed: thawing *(drip loss) and cooking (cooking loss),* which showed the following values for each process (Figure 17 and Appendices G and H).

Defrosting : showed a loss of between 3.06 and 7.31% with an average of 5.41% (±1.30).

▪*Coction* : expressed a range of loss from 5.31 to 38.35%, with an average of 21.22% (± 9.72)

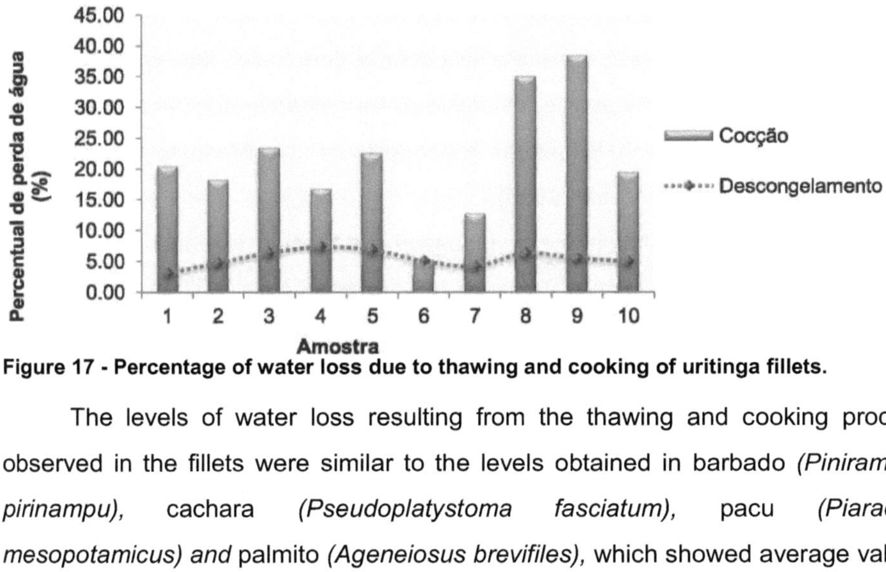

Figure 17 - Percentage of water loss due to thawing and cooking of uritinga fillets.

The levels of water loss resulting from the thawing and cooking process observed in the fillets were similar to the levels obtained in barbado *(Pinirampus pirinampu),* cachara *(Pseudoplatystoma fasciatum),* pacu *(Piaractus mesopotamicus) and* palmito *(Ageneiosus brevifiles),* which showed average values according to Table 09 (CASTRO, 2007). It is worth noting that the values obtained from the cooking process alone were similar to the levels of water loss found in representatives of the mollusca phylum, such as *Octopus vulgaris* (39.75 - 41.64%) and *Octopus insularis* (39.07 - 40.12%) (SILVA, 2009).

Table 09. Average percentage values of water loss due to defrosting and cooking of some fish species.

VULGAR NAME	SCIENTIFIC NAME	WATER LOSS (%)	
		DEFROSTING	COCTION
Barbado	*P. pirinampu*	**5,15**	**76,03**
Cachara	*P. fasciatum*	**5,19**	**76,77**
Pacu	*P. mesopotamicus*	**3,36**	**83,50**
Palm hearts	*A. brevifiles*	**3,07**	**82,77**

The cooking process showed the highest levels of water loss in the muscle, with an average value of 21.22%, which represents a reduction in the weight of the fillet of 13.84 g on average, when correlated with the percentage of water loss due to thawing, which expressed 5.41%, pressurising a reduction in the weight of 3.87 g of the fillet. It can also be seen that the fillets subjected to the cooking process had a 15.18% greater water loss than those subjected to the thawing process (Table 10).

Table 10 - Average water loss values of fillets subjected to the thawing and cooking process.

TYPE OF PROCESS	INITIAL WEIGHT (g)	FINAL WEIGHT (g)	WEIGHT LOSS (g)	% WATER LOSS
Defrosting	**61, 87**	**58,48**	**3,87**	**5,41**
Cooking	**64,22**	**50,38**	**13,84**	**21,22**

A similar situation found in the uritinga fillets with regard to the process that caused the greatest water loss was also described by Castro (2007) when he studied water loss due to defrosting, cooking and pressing in barbado, cachara, pacu and palmito fillets, in which he found that the cooking process caused between 76.03 and 82.77% more water loss than the defrosting process.

In view of the above, it can be seen that a significant difference in the percentage of water loss resulting from the cooking process may be due to the influence of the different temperature exposure during heating of active sites in the hydrophilic domains of muscle proteins, since these proteins are the most affected by heat and tend to lose more water (OFSTAD et al. 1996 b ; CASTRO, 2007).

When the percentage of water loss in the fillets was compared with the pH values, regardless of the process used, water loss showed no direct influence on pH (Figure 18). This relationship was contradictory to the information provided by Castro (2007) when he mentioned that water loss is directly related to pH, where the higher the pH value there is a tendency for less protein denaturation and consequently greater water retention capacity.

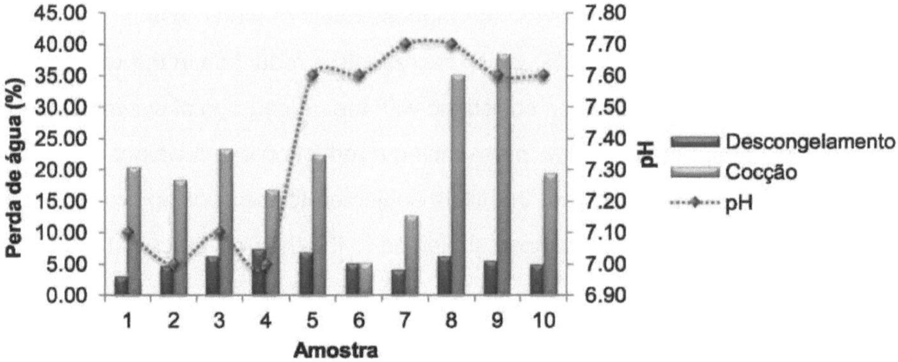

Figure 18. Relationship between water loss due to thawing and cooking and pH of uritinga fillets.

The parameter in question, water loss, is a property of fundamental importance in terms of the quality of the product destined for industrialisation and consumer acceptability. This property is based on the behaviour of the meat when it is subjected to the freezing, thawing and cooking process, which in turn influences its water retention capacity (SOUZA, 2006; SILVA, 2009).

Water loss has an interdependent relationship with raw material characteristics such as morphometric features (body length), weight, centesimal composition, gonadal development, etc., and especially pH, due to its influence on protein denaturation levels and, consequently, on the retention capacity and percentage of water loss in muscle tissue (OFSTAD et al, 1995b; ROMAN and SGRABIERI, 2005).

It is important to note that water loss due to thawing and cooking is the result of chemical, physical-chemical and structural alterations to the compartmentalisation components of the cell membrane, reduction in muscle fibres, protein coagulation, mybiofibrillar and sarcoplasmic aggregation and reduction and solubilisation of connective and adipose tissue, causing dehydration and reduction in water retention capacity (LAAKKONEN et al, 1970; FAUCONNEAU et al, 1995; OFSTAD et al.,1996 b; CALIFANO et al, 1997; CARECHE et al, 1999; GARCIA et al, 1999; TSCHEUSCHNER, 2001; TORNBERG, 2005).

The intensity of these changes varies according to the process the muscle is subjected to, for example, thawing, visualised by dripping, varies according to the species, the cooling and freezing method and the storage time and temperature (BELLO et al, 1982; CARECHE et al, 1999; GARCÍA et al, 1999), while cooking varies according to the form of heat transfer, the temperature, the duration of the process and the cooking medium (POTTER and HOTCHKISS, 1995; GARCIA-ARIAS et al., 2003).

6.4.3 Water Retention Capacity (WRC)

Water retention capacity (WRC) is one of the most important functional properties when it comes to preparing and adding value to meat products and increasing consumer acceptance of the quality attributes of fillets. This property influences the appearance, palatability, reduction in size and juiciness of processed products and is directly related to the capacity of the muscle to lose liquid before and during the thawing, cooking and pressing processes (BRESSAN, 1998; ANADÓN, 2002; CASTRO, 2007; LARA et aL, 2010).

In the present study, the water retention capacity observed in uritinga fillets showed an average value of 51.55 % (± 2.56) with a range from 46.36 to 55.11 % (Figure 19 and Appendices G and H).

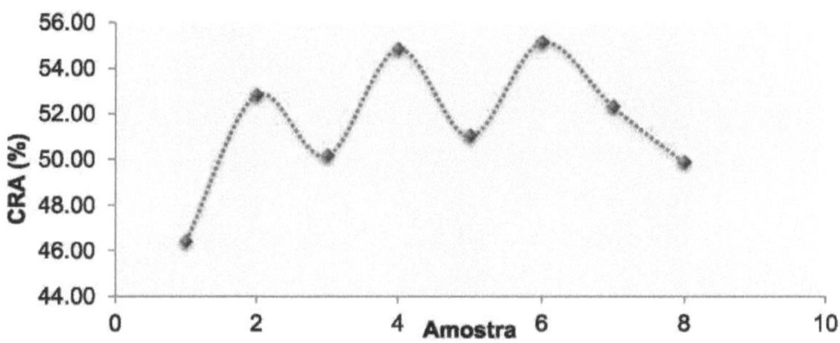

Figure 19 - Water retention capacity (WRC) of uritinga fillets.

These values were lower than those found in fillets obtained from other fish species such as surubim (LARA et al, 2010) and salmon (OFSTAD et al., 1995), and when compared with the percentage obtained for barbado, cachara, pacu and palmito, according to Castro (2007), these values were higher.

According to Castro (2007), the range of variation between the maximum and minimum value for CRA may be related to various factors, such as stress and handling during fishing and the genetic and nutritional characteristics of the raw material.

CHAPTER 7

CONCLUSION

- Uritinga has a fillet yield compatible with fish species considered noble

- The data on morphometric characteristics, fillet yield and centesimal composition associated with physico-chemical parameters (CRA, pH and water loss) show that the species has good potential for industrialisation.

- In order to use the species in the fish industry, further studies are needed on when the species is most and least available, the best processing techniques that can be used, and verification of economic viability^, among other factors.

REFERENCES

ALBERT, C.M.; HENNEKENS, C.H.; O'DONNELL, C.J.; AJANI, U.A.; CAREY, V.J.; WILLETT, W.C.; RUSKIN, J.N.; MANSON, J.E. Fish consumption and risk ofsudden cardiacdeth. JAMA, v.279, n.01, p.23-28, 1998.

ALCALÁ, C. A. L.; SÁNCHEZ-DUARTE, P. **The Orinoco delta fish:** Diversity, bioecology, use and conservation. Venezuela: Fundaci6n Salle de Ciencias Naturales, 2011.504 p.

ALMEIDA, M.C. **Body relations, centesimal composition and fillet yield of mandi *(Pimelodus brítskii),* from the Salto Santiago Rio Iguaçu reservoir.** 2010. 44 f.Dissertation (Master's in Animal Science), State University of Western Paraná, Cândido Rondon, 2010.

ALMEIDA, N.M. **Post-mortem changes in *Colossoma macroporun* (CUVIER, 1818) from fish farming and preserved on ice.** 1998.90f. Dissertation (Master's Degree in Food Sciences), University of Amazonas, Manaus, 1988.

ALMEIDA, Z. S. **The marine and estuarine fishing resources of Maranhão:** Biology, Technology, Socio-economics, State of the Art and Management. 2008. 283f. Thesis (Doctorate in Zoology), Federal University of Pará, Atheneu, 2008.

ALVES, C.C.; WAITZBERG, D. Proteins. *In:* WAITZBERG, D.L.(Ed.). **Oral, enteral and parenteral nutrition in clinical practice.** 4.ed.São Paulo: Atheneu, 2009. p . 85-108.

ANADÓN, H. L. S. **Biological, nutritional and Processing factors affecting breast meat quality of broilers.** 2002.171 f. Thesis (Doctor of Philosophy in Animal and Poultry Sciences), Faculty of Virginia Polytechnic Institute and State University, 2002.

AOAC - ASSOCIATION OF OFFICIAL ANALYTICAL CHEMISTS. **Official methods of analysis of AOAC International.** 17 Ed. Gaithersburg: AOAC International, 2000. 937 p.

ARAÚJO, F.G. Relative abundance, distribution and seasonal movements of marine catfish (Siluriformes, Ariidae) in the estuary of Lagoa dos Patos, RS. **Revista Brasileira de Zoologia,** Curitiba, v. 5, n. 4, p. 509-543, 1988.

AZEVEDO, J.W.J. ; CASTRO, A.C.L.Weight-length relationship and condition factor of the uritinga, *Hexanematichthys proops,* (Valenciennes, 1840) (siluriformes, ariidae), caught on the western coast of Maranhão. **Boi. Lab. de Hidrobiologia,** n. 21, p.75-82, 2008

BANDARRA, N. M, BATISTA, I.; TAFULA, J.; NUNES, M. L. Nutritional Value of Fishery Products. In: IPIMAR. Fisheries Products: Quality, Safety and Technological Innovation. **Actas das Jornadas Técnicas e Cientificas do IPIMAR,** p.75-87, 2001.

BELLO, R. A.; LUFT, J. H.; PIGOTT, G. M. Ultrastructural study of skeletal fish muscle after freezing at different rates. **Journal of Food Science,** Chicago, v. 47, n. 5, p. 1389-1394, 1982.

BERNARDES, R. A.; FIGUEIREDO, J. L.; RODRIGUES, A. R.; FISCHER, L. G.; VOOREN, C. M.; HAIMOVICI, M.; ROSSI-WONGTSCHOWSKI, C. L. D. B. **Fishes of the Exclusive Economic Zone of the southeast-south region of Brazil: a** survey using traps, hawksbills and bottom trawls. São Paulo, Edusp, 2005. 295p.

BITTENCOURT, F. **Cultivation of pacu *Piaractus mesopotamicus at* different densities in net pens in the Itaipu Reservoir.** 2008.46 f. Dissertation (Master's in Zootechnics), State University of Western Paraná, Paraná, 2008.

BLACKITII, R.E.; REYMENT, R.A. **Multivariate Morphometry.** London: Academic Press, 1971.410 p.

BOCHI, V.C.; WEBER, J.; RIBEIRO, C. P.; VICTÓRIO, M. A.; EMANUELLI, T. Fishburger with silver catfish *(Rhamdia quelen)* filleting residue. **Bioresource Technology,** v. 99, p. 8844-8849, 2008.

BOMBARDELLI, R. A.; SANCHES, E. A. Evaluation of body morphometric characteristics, cut yield and centesimal composition

of the meat of the armado *(Pterodoras granulosus).* **Boletim do. Instituto da Pesca,** São Paulo, v. 34, n.2, p. 221 - 229, 2008.

BOMBARDELLI, R. A.; SYPERRECK, M. A.; SANCHES E. A. Current situation and prospects for consumption, processing and adding value to fish. **Arquivos de Ciências Veterinárias e Zoologia,** Umuarama, v.8, n.2, p181-195, 2005.

BOOKSTEIN, FL.; CHERNOFF, B.; ELDER, R.; HUMPHRIES, J.; SMITH, G.; STRAUSS, R. Morphometrics in evolutionary biology. **Acad. ato Sei,** Philad, n.15, p. 1-277, 1985.

BOSWORTH, B.G.; LIBEY, G.S.; NOTTER; D.R. Relationship among total weight, body shape, visceral components, and fillet traits in palmetto bass (Striped bass female *Morone saxatilis* x white bass male *M. chrysops)* and paradise bass (Striped bass female *Morone saxatilis* x yellow bass male *M. mississippiensis).* **Journal of the World Aquaculture Society,** Los Angeles, v. 29, n.1,p 40-50, 1998.

BRAZIL. Ministry of Agriculture, Fisheries and Supply. **Ordinance no. 185, of 13 May 1997:** Technical regulation on the identity and quality of fresh fish (whole and gutted). Diário Oficial da União, Brasília, n.93, 19 May 1997. Section I, p.10. 282-283.

BRAZIL. Ministry of Agriculture, Fisheries and Supply. **Regulations for the Industrial and Sanitary Inspection of Products of Animal Origin.** Approved by Decree No. 1.255, of 29 March 1962. Brasília: MAPA, 1980.p.74-76.

BRESSAN, M. C. **Effect of pre- and post-slaughter factors on the quality of chicken breast meat. 1998.**201 f. Thesis (Doctorate in Food Technology), Faculty of Food Engineering, State University of Campinas, Campinas, 1998.

BRESSAN, M. C. **Effect of pre- and post-slaughter factors on the quality of chicken breast meat. 1998.**201 f. Thesis (Doctorate in Food Technology), Faculty of Food Engineering, State University of Campinas, Campinas, 1998.

BRITO, T.P. **Survey of past data and analysis of information on the biology and fishing of estuarine fish in the Northern Region of Brazil.** 2005. 84 f. Final Course Work (Undergraduate Degree in Oceanography), Federal University of Para, Belém, 2005.

BRUSCHI, F. L. F. **Yield, chemical composition and fatty acid profile of fish and fish waste:** a comparison. 2001.65 f. Monograph (Degree in Oceanography), Universidade do Vale do Itajaí, Itajaí, 2001

BRUSCHI, F.L.F. **Yield, chemical composition and fatty acid profile of fish and fish waste: a comparison.** 2001. 65f. Monograph (Final Paper for the Degree of Oceanographer) - Oceanography Course, Universidade do Vale do Itajaí, 2001.

BUCKLEY, J.T.; GROVES, T.D.D. Influence of feed on the body composition of finfish. In: SYMPOSIUM ON FINFISH NUTRITION AND FEED TECHNOLOGY, EIFAC/78/SYMP:R/17,Hamburg, 1978. Proceedings... Hamburg: 14p.

CALIFANO, A. N. et al. Effect of Processing conditions on the hardness of cooked beef. **Journal of Food Engineering,** Oxford, v. 34, n. 1, p. 41-54, 1997.

CAMARGO, M.; ISAAC, V. J. N. Estuarine fishes of the northern region of Brazil: List of species and considerations on their geographical distribution. **Boletim do Museu Paraense Emílio Goeldi,** v. 17, n. 2. p. 133 - 157, 2001.

CANTANHÊDE, G.; DE CASTRO, A.C.L.; GUBIANI, E.A. Reproductive biology of *Hexanematichthys proops* (Siluriformes, Ariidae) on the western coast of Maranhão. **Iheringia, Sér. Zool.,** Porto Alegre, v. 97, n. 4, p. 498-504, 2007

CARACIOLO, M. S. B.; KRUGER, S. R.; COSTA, F. J. C. B. Filleting strategies and utilisation of tambaqui meat. **Panorama da Aquicultura,** Rio de Janeiro, v.11, n.67, p. 25 - 29, 2001.

CARECHE, M. et al. Structural changes of hake *(Meriuccius meiluccius* L.) fillets: effects of freezing and frozen storage. **Journal of Ag ri cultural and Food Chemistry,** Washington, v. 47, n. 3, p. 952-959, 1999

CASTRO, D. A. **Water losses in pantanal fish fillets.** 2007. 50f. Dissertation

(Master's in Animal Science), Federal University of Mato Grosso do Sul, Campos Grande, 2007.

CERVIGÓN, F. **Los peces marinos de Venezuela,** v.1, 2.ed. Caracas: Fundación Científica Los Roques, 1991.425 pp.

CEZARINI, R. **Adding value to freshwater fish**: smoking jundiá (R/?an7d/a *quelen)* fillets. 2005. 52f. Final course work (Degree in Food Engineering), Universidade do Vale do Rio dos Sinos, São Leopoldos, 2005.

CHEFTEL, J. C.; CHEFTEL, H. **Introducción a la bioquímica y tecnologia delos alimentos.** Spain: Acribia, 1976, p. 333.

CIBERT, C.; FERMON, Y., VALLOD, D.; MEUNIER, F.J. Morphological screening of carp *Cyprinus carpio:* relationship between morphology and fillet yield. **Aquatic Living Resources,** v.12 , n.1, p.01-10, 1999

COELHO, G.M. Yield and chemical composition of fish and residual carcasses from industrial filleting: a comparison. /n:BRAZILIAN WORKSHOP ON THE USE OF FISH SUB-PRODUCTS, 1, 2003, Itajaí (SC), **Electronic Proceedings...**Itajaí: Univali, 2003. CD-ROM.

CONNOR, S.L.; CONNOR, W.E. Are fish oils beneficial in the prevention and treatment of coronary disease? **American Journal of Clinical Nutrition,** v.66, n.4, p.1020-1031, 1997.

CONTRERAS-GUZMÁN, E. **Biochemistry of fish and fish products.** Jaboticabal: FUNEP. 1994.409p.

CORDEIRO, L. G. M. S. **Este róis como marcadores moleculares da contaminação fecal no sistema estuarino Iguaçu-Sarapuí, noroeste da Baía de Guanabara** (RJ).2006.170 f.Dissertação(Mestrado em Química Analítica), Universidade Católica do Rio de Janeiro, Rio de Janeiro, 2006

CORSON, W. H. **Global ecology handbook: what you can do about the environmental crisis.** São Paulo: Augustus, 2002. 413 p

CREPALDI, D.V.; FARIA, P.M.C.; TEIXEIRA, E. A.; RIBEIRO, L. P.; COSTA, A. A. P.; MELO, D. C.; CINTRA, A. P. R.; PRADO, S. A. ; COSTA, F. A. A.; DRUMOND, M. L.; LOPES, V. E.; MORAES, V. Surubim in Brazilian aquaculture. **Rev Bras**

Reprod Anim, Belo Horizonte, v.30, n.3/4, p.150-158, jul./dez. 2006.

CYRINO, J.E.; URBINATI, E.C.; FRACALOSSI, D.M.; CASTAGNOLLI, N. **Tópicos especiais em piscicultura de água doce tropical intensiva.** São Paulo: Brazilian Society of Aquaculture and Aquatic Biology, 2004. 533p.

DAMASCENO, A. **Quality (sensory, microbiological, physico-chemical and parasitological) of chilled salmon *(Salmo salar,* Linnaeus, 1778) sold in Belo Horizonte - MG .** 2009.48f. Dissertation (Master's Degree in Animal Science in the area of Technology and Inspection of Products of Animal Origin), Federal University of Minas Gerais, Veterinary School, Belo Horizonte, 2009.

ETCHEVERS, S.L. Contribution to the biology of the sea catfish, *Arius spixii* (Agassiz) (Pisces- Ariidae), South of Margarita Island, Venezuela. **Bui. Marine of Science.** 1978. v. 28, p. 381-385.

EVANGELISTA, J.P. **Food Technology.** 2ª Ed. São Paulo: Atheneu,1998. p.313-315. 1998

EYO, A.A. **Carcass composition and filleting yield of ten fish species from Kainji Lake.** Proceedings of the FAO expert consultation on fish technology in Africa, Ghana, 22-25 October, 1991. FAO: Fish. Rep., suppL, n.467,p 73- 175, 1993.

FALABELLA, P.G. **Fishing in the Amazon River:** problems and Solutions. Manaus: Edua, 1994.156 p.

FALCH, E.; OVERREIN, I.; SOLBERG, C.; SLIZYTE, R.Composition and calories. /n:NOLLET, L.M.L.; TOLDRÁ (Eds.). **Handbook of seafood and seafood products analysis.** BocaRaton: CRC Press - Taylor & Francis Group, 2010. p .257-285.

FAO - Food and Agriculture Organisation of the United Nations. **The State of World Fisheries and Aquaculture,** 2012. Available at: <http://www.fao.org/fishery/sofia/en>. Accessed on: 19 August 2012.

FAO. **The State of world fisheries and aquaculture 2006.** Fisheries Technical Document. 2007. Available at: <http://www.fao.org>. Accessed on: 26 August 2012

FARIA, R. H. S.; SOUZA, M. L. R.; WAGNER, P. M.; POVH, J. A.; RIBEIRO, R. P. Processing yield of Nile tilapia *(Oreochromis niloticus* Linnaeus, 1757) and pacu

(*Piaractus mesopotamicus* Holmberg, 1887). **Acta Scientiarum, Animal Sciences,** Maringá, v. 25, n. 1, p. 21-24, 2003.

FAUCONNEAU, B. et al. Growth and meat quality relations in carp. **Aquaculture,** Amsterdam, v. 129, n. 1-4, p. 265-297, 1995.

FÁVARO, L.F.; FREHSE, F.A.; OLIVEIRA, R.N.; JÚNIOR, R.S. Reproduction of the yellow *catfish*, *Cathorops spixii* (Siluriformes, Ariidae), from Pinheiros Bay, estuarine region of the Paraná coast, Brazil. **Rev. Bras. Zoologia,** v. 22, n. 4, p.1022-1029, 2005

FENNEMA, R.O. **Química de los Alimentos.** 3° ed. Zaragoza, Spain: Acríbia, 1258p. 2000.

FERREIRA, L. K. S.; EVERTON, F. A.; FRAZÃO, F.B; PEREIRA, T. J.F. Characterisation of marine fisheries production at the Portinho Landing Station, São Luís-MA. *In:* SEMANA NACIONAL DE OCEANOGRAFIA, 23,2011, São Luís, Anais...São Luís: UFMA, 2011. CD- ROM

FERREIRA, L. K. S.; PEREIRA, T. J. F.; LIMA, M. F. V.; RIBEIRO, A. C. Analysis of the hygiene and sanitary aspects of fish sold at the fish market in São Luís - MA. *In:* CONGRESSO BRASILEIRO DE ENGENHARIA DE PESCA, 16, 2009, Natal, Anais...Natal:ABEP, 2009 .CD- ROM.

FERREIRA, M.W.; SILVA, V. K.; BRESSAN, M.C.; FARIA, P. B.; VIEIRA, J. O.; ODA, S. H. I. **Processed fish:** longer shelf life and greater added value. Lavras: Federal University of Lavras, 2002. 26 p.

FERRETTI, R.; DUARTE, R.A.; TERRA, N.L.; MORIGUCHI, Y. Atherosclerosis and omega-3 fatty acids. **Acta Méd.,** Porto Alegre, n.15, p.557-574,1994.

FIGUEIREDO, J. L.; MENEZES, N. A. **Manual of marine fish from south-east Brazil. II.** Teleostei (1). São Paulo, Zoology Museum of the University of São Paulo, 1978. 110p.

FLORES, J.; BERMELL, S. Functional properties of myofibrillar proteins: water retention capacity. **Revista de Agroquímica e Tecnologia de Alimentos,** v.24, n.2, p.151-158,1984.

FOGAÇA, F. H.S.; LEGAT, A. P.; PEREIRA, A. M. L.; LEGAT, J. F. A. **Methods for analysing fish.** Teresina: Embrapa Meio-Norte, 2009.40 p. (Série Documentos da Embrapa Meio Norte).

FOOD INGREDIENTS BRASIL. Fish proteins: functional properties of fish proteins. **Food Ingredients Brasil Magazine,** São Paulo, n 08, p 23-32, 2009.

FREATO.T. A.; SANTOS, V. B.; RILKE FREITAS, T. F.; REIS NETO, R. V.; ALLAMAN, I. B.; LAGO, A. A.; PIMENTA, M. E.S. G.; COSTA, A. C. Morphometric evaluation of the piracanjuba *Brycon orbignyanus* as a function of body weight . *In*:CONGRESSO BRASILEIRO DE PRODUÇÃO DE PEIXES NATIVOS DE ÁGUA DO DOCE, 1, ENCONTRO DE PISCICULTURORES DE MATO GROSSO DO SUL,1, Dourados(MS), 2007, **Anais eletrónicos...**Campo Grande: Embrapa, 2007(CD-ROM).

FREATO, T.A. **Morphometry, processing yield and interrelationships in the evaluation of Piracanjuba carcasses, *Brycon orbignyanus* (VALENCIENNES,1849).** Lavras, 2005. 90p. Dissertation (Master's Degree in Zootechnics), Federal University of Lavras, Lavras, 2005.

FREITAS, J.V.F.; GURGEL, J.J.S. Study of some biometric parameters and chemical composition of tambaqui, *Colossoma macroporum* (CUVIER, 1818) and pirapitinga *Colossoma brachyporum* (CUVIER, 1818), raised in captivity. DNOCS **Technical Bulletin**, v. 43, n.1, p.47-66, 1985.

GAGLEAZZI, U. A.; GARCIA, F. T.; BLISKA, F. M. M. Characterisation of meat consumption in Brazil. **Revista Nacional Da Carne,** São Paulo, v.26 n.310, p. 152-160, 2002.

GARCÍA, M. L. et al. Ultrastructure of the myofibrillar component in cod *(Gadus mortiua* L.)and hake *(Meriuccius meriuccius* L.) stored at -20°C as a function of time. **Journal of Agricultural and Food Chemistry,** Washington, v. 47, n. 9, p. 3809-3815, Sept. 1999.

GARCIA-ARIAS, M. T. et al. Cooking-freezing-reheating (CFR) of sardine *(Sardina pilchardus)* fillets: effect of different cooking and reheating procedures on the proximate and fatty acid compositions. **Food Chemistry,** Oxford, v. 83, n. 3, p. 349-356, 2003.

GARUTTI, V. Revalidation of *Astyanax rupununi* Fowler, 1914 (Teleostei.Characidae) and description of two new species for the genus. **Papéis Avulsos de Zoologia,** Museu de Zoologia da Universidade de São Paulo, São Paulo, v.43, n.1, p.1.9,2003

GERMANO, P.M.L.; GERMANO, M.I.S.; OLIVEIRA, C.A.F. Aspectos da qualidade do pescado de relevância em saúde pública. **Food Hygiene,** v.12, n.53, p.30-37, 1998.

GOMIERO, J. S. G.; RIBEIRO, P. A. P.; FERREIRA, M. W.; LOGATO, P. V. R. Carcass yield of matrinxã fish *(Brycon cephalus)* in the different cuts of the head. **Ciência Agrotecnologica,** Lavras ,v. 27, n.1, p.211-216, 2003.

GONÇALVES, A. A. General Aspects of Fish. *In:* GONÇALVES, **A.A.(Ed).Tecnologia do pescado:** ciência, tecnologia, inovação e legislação.São Paulo:Atheneu, 2011 .p. 02 - 09.

GONÇALVES, A. A.; PASSOS, M. G. Restructured Fish Product from White Croacker *(Micropogonias furnierí)* Mince Using Microbial Transglutaminase. **Brazilian Archives of Biology and Technology,** Brazil, v.53, n. 4, p. 987- 995, jul/ago, 2010

GONÇALVES, A.A. General Aspects of Fish. In: GONÇALVES, **A.A.(Ed).Tecnologia do pescado:** ciência, tecnologia, inovação e legislação.São Paulo:Atheneu, 2011 .p. 02 - 09

GONÇALVES, A.A.; MENEGASSI, M. Chemical composition and calorific value of the main Brazilian fish products. *In:* GONÇALVES, A.A.(Ed).Tecnologia **do pescado:** ciência, tecnologia, inovação e legislação.São Paulo: Atheneu, 2011.p. 573-581.

GOODMAN, R.K.A. **A comparison of morphometric characteristics of channel catfish,** *Ictalurus punctatus* **(Rafinesque) from seven different geographical locations.** Auburn,1973. 27p. Thesis (M.S), Aubum University, Auburn,1973.

GUEDES, D.S.; VASCONCELOS FILHO, A.L. Ecological Study of the Itamaracá Region, Pernambuco, Brazil. IX. Information on the Feeding of White and Yellow Catfish (Pisces, Ariidae). **Oceanographic Work of the Federal University of Pernambuco,** Recife, v. 15, p. 323 - 330, 1980.

GUHA, B. C., The Role of Fish in Human Nutrition; **Fish in Nutrition.** Edited by E. Heen and R. Kreuzer; FAO, Rome - Italy; pp 39 - 42; 1962.

HONIKEL, K.O.; HAMM, R. Measurement of water-holding capacity and juiciness. *In:* PEARSON, A.M.; DUTSON, T.R. (Eds). **Quality attributes and their measurement in meat, Poultry and fish products.** United States: Kluwer Academic, 1994. p. 125-159.

HUANG, C.M.; LIAO, LC. Response to mass selection for growth rate in *Oreochromis nlloticus.* **Aquaculture,** Amsterdam, v. 85, n. 1/4, p. 199-205, mar. 1990.

HURLBUT, T.; CLAY, D. 1998 Morphometric and meristic differences between shallow and deep-water populations of white hake *(Urophycis tenuis)* in the Southern Gulf of St. Lawrence. *Can. J. ofAni. Sei.,* Toronto, n.55,p 2274-2282.

HUSS, H. H. **Fresh fish:** its quality and quality changes. Rome: Food and Agriculture Organisation (FAO) of the United Nations. 1988. 132p.

IABS - Sustainable Brazil Environmental Institute. **Fishing complex in São Luís:** Version for Discussion. Brasília: IABS, 2008.111 p.

IBAMA/SEAP/PROZEE, 2008. **Monitoring fishing activity on the northeastern coast - Estatpesca Project.** Tamandaré, 385p.

JACQUOT, R. Organic constituents of fish and food. *In:* **Fish and food.** New York (USA): G. Borgsrom, 1961. p. 144-192.

JONSSON, A.; SIGURGISLADOTTIR, S.; HAFSTEINSSON, H.; KRISTBERGSSON, K. Textural propertiesof raw Atlantic salmon *(Salmo salar)* fillets measured by different methods in comparison to expressible moisture. ***Aquaculture Nutrition,*** n.81, p. 81-89, 2001.

JUSTI, K.C.; HAYASHI, C; VISENTAINER, J.V.; SOUZA, N.E.; MATSUSHITA, M. The influence of feed supply time on the fatty acid profile of nile tilapia *(Oreochromis niloticus)* fed on a diet enriched with n-3 fatty acids. **Food Chemistry,** v.80, p.489-493, 2003.

KIETZMANN, U.; PRIEBE, K.; RAKOW, D.; REICHSTEIN, K. Inspection of fish, crustaceans and molluscs. In:. **Veterinary inspection of fish.** Zaragoza, Spain:

Acribia, 1974. 326 p. ch. 5, p. 147-240

KIM, J.D.; LALL, S.P.Amino acid composition of whole body tissue of Atlantic halibut *(Hippoglossus hippoglossus),* yellowtail *flounder(Pleuronectes ferruginea)* and Japanese flounder *(Parallchthys oliíaceus).* **Aquaculture,** n. 187, p. 367-373, 2000.

KNOPPERS, B.; SOUZA, W.F.L.; EKAU, W.; FIGUEIREDO,A.G.; SOARES-GOMES, A. The Land-Sea Interface of Brazil. *In:* PEREIRA,R.C. & SOARES-GOMES, A.(Org).**Biologia Marinha.**2ª Ed.Rio de Janeiro:Interciência,2009. p. 529-553.

KUBITZA, F. Coletânea de informações aplicadas ao cultivo do tambaqui, do pacu e de outros peixes redondos. **Panorama da Aquicultura,** v.14,n.83, p 13-23, mai/jun, 2004

KUBOTA, E.H.; EMANUELLI, T. Fish processing. In: BALDISSEROTTO, B.; RADÚNZ, J. **Criação de Jundiá.** Santa Maria (RS): Ed. UFSM, 2004. 232p.

KUHN, C. R., SOARES, G. J. D. Proteases and inhibitors in surimi processing. **Revista Brasileira de Agrociência,** Pelotas, v. 8, n. 1, p. 5-11, jan./abr. 2002.

KUWABARA, W.T.;HILSDORF, A. W. S.; MARCENIUK, A. P. Taxonomic identity of species of the genus *Astyanax* Baird & Girard, 1854 (Characiformes; Characidae), from the headwaters of the Tietê River, São Paulo. In: Congresso de iniciação cientifica da Universidade de Mogi das Cruzes, 14, Mogi das Cruzes (SP),2011, **Annuais eletrónicos...** Mogi das Cruzes: UMC, 2011 (CD-ROM)

LAAKKONEN, E.; WELLINGTON, G. H.; SHERBON, J. W. Low temperature, long-time heating of bovine muscle. 1. Changes in tenderness, water-binding capacity, pH and amount of water-soluble components. **Journal of Food Science,** Chicago, v. 35, n. 2, p. 175-177, 1970.

LAKSHMANAN, R.; PARKINSON, J. A.; PIGGOTT, J. R. High-pressure processing and water-holding capacity of fresh and cold-smoked salmon *(Salmo salar).* **Lebensmittel- Wissenschaft und- Technologie,** v. 40, p. 544 -551,2007.

LANARI, D.; POLI, B.M.; BALLESTRAZI, R. D'AGARO E.; MECATTI M. The effects of dietary fat and NFE levels on growing european sea bass *(Dicentrarchus labraxL.).* Growth rate, body and fillet composition, carcass traits and nutrient retention efficiency. **Aquaculture,** v.179, p.351-364, 1999.

LARA, J. A. F.; GARBELINI, J. S.; DELBEM, Á. C. B. Determination of the Water Retention Capacity of Pintado Fillets from the Paraguay River (Corumbá-MS). *In:* SIMPÓSIO SOBRE RECURSOS NATURAIS E SOCIOECONÔMICOS DO PANTANAL, 5,2010, Corumbá (MS),Anais **eletrónicos...**Corumbá:EMBRAPA , 2010.4p.(CD-ROM).

LE BAIL, P.-Y.; KEITH, P.; PLANQUETTE, P. Atlas des poissons d'eau douce de Guyane (tome 2, fascicule II). **Publications scientifiques du Muséum national d'Histoire naturelle,** Paris. 2000. 307 p.

LEDENER, J. **Modern encyclopaedia of food hygiene.** São Paulo: Manole, v.2, 1991

LEHNINGER, A.L.; NELSON, L.D.; COX, M.M. **Principles of Biochemistry.** 2006. 4 ed.São Paulo: Editora SARVIER. 1232 p.

LEIS, J. M.; TRNSKI.T. **The larvae of Indo-Pacific shorefishers.** Sydney: University of Hawaii Press; The Australian Museum, 371 p. 1989.

LEISTNER, L.; GORRIS, L.G.M. Food preservation by hurdle technology. **Trends in Food Science & Technology,** v.6, p.41-45, feb.1995.

LEONHARDT, J. H.; CAETANO FILHO,M.; FROSSARD; H.; MORENO, A.M. Morphometric characteristics, yield and composition of the fillet of Nile tilapia, *Oreochromis niloticus, of* the Thai, local and crossbred strains. **Semina:** Ciências Agrárias, Londrina, v. 27, n. 1, p. 125-132, jan./mar. 2006.

LIMA, M. F. V. **Physiological anatomical aspects of fish of industrial importance.** São Luís: UEMA, 2002.20p.

LOPES, P.R.S.; POUEY, J.L.O.F.; ENKE, D.B.S.. Performance of *Rhamdia quelen fingerlings* fed different levels of energy in the diet. **Revista Biodiversidade Pampeana,** v.4, n.13, p.32-37, 2006.

LOPES, S. **Biology.** São Paulo: Saraiva, 1999.607p.

MACEDO-VIEGAS ,E.M.;SOUZA, M.L.R.Pre-processing and conservation of farmed fish *In:* CYPRINO, J.E.P.; (Ed.); URBINATI, E.C.; PEREIRA, **C.R. Tópicos especiais em piscicultura de água doce tropical intensiva.** São Paulo: TecArt, 2004. p.405-480.

MACEDO-VIEGAS, E. M.; FRASCÁ-SCORVO, C. M. D.; VIDOTTE, R. M.; SECCO, E. M. Effect of weight classes on the body composition and processing yield of farmed matrinxã *(Brycon cephalus).***Acta Scientiarium,**Maringá, Paraná, v. 22, n. 3, p. 729-732, 2000.

MACEDO-VIEGAS, E. M.; SOUSA, M. L. R.; KRONKA, S. N. Study of the carcass of Nile Tilapia *(Oreochromis niloticus)* in four weight categories. **Revista UNIMAR,** Maringá, v. 19, n. 3, p. 863-870, 1997.

MACHADO, M.R.F.; FORESTI, F. Yield and chemical composition of *Prochilodus lineatus* fillet from the Mogi Guaçu River, Brazil. **Archivos de Zootecnia,** Córdoba, v. 58, n.224, p. 663-670, 2009.

MACHADO, Z. L. Chemical composition of fish. In: MACHADO, Z. L. **Tecnologia de recursos pesqueiros, parâmetros, processos, produtos.** Recife: DAS/DA, 1984.

Machado, Z.L. **Tecnologia de recursos pesqueiros:** parâmetros, processos, produtos. Recife: Sudene-DRN-Div.Recursos Pesqueiros, 1984. 277p.

MACKIE, LM. Fish protein. *In:* HUBSON, B.J.G.(Ed.). **New and developing sources of food proteins,** New York: Chapman & Hall, 1994, p. 95-137.

MAHAN, B.M.; MYERS, R. J. Biochemistry.TOMA, E. H(Coord). **Chemistry:** a university course. Trad. ARAKI, K.; SILVA, D.O.; MATSUMOTO, F. M.São Paulo: Edgard Blucher, 1995.582p.

MEDEIROS, S. D. **Technology and inspection of fish and** derivatives: chemical composition of fish, São Paulo: Qualitas, 2012.10p.

MENDES, A. A.; MOREIRA, J.; GARCIA, R.G. Quality of broiler breast meat. **Revista Nacional da Carne,** São Paulo, v.27, n.317, p.138- 144, 2003.

MENEGASSI, M. Nutritional Aspects of Fish. In: GONÇALVES, **A.A.(Ed).Tecnologia do pescado:** ciência, tecnologia, inovação e legislação.São Paulo:Atheneu, 2011 .cap 1.5.p. 43 - 60

MINOZZO, M. G. **Preparation of creamy pâté from Nile tilapia fillet *(Oreochromis niloticus)* and its physico-chemical, microbiological and sensory characterisation.** 2005. 127 f. Dissertation (Master's in Food Technology) - Federal University of Paraná, Curitiba, 2005.

MIRANDA,M.O.T.;RIBEIRO,L.P.Zootechnical characteristics of the surubim *(Pseudoplastiloma coruscans). In:* MIRANDA, M.O.T (Org.) **Surubim.** Belo Horizonte: IBAMA, 1997. p.43-56 (Coleção Meio-Ambiente, Série Estudos Pesca, 19).

MORAES, D. A. Geometric morphometrics and the "revolution in morphometrics" locating and visualising changes in the shape of organisms. **Bioletim,** Rio de Janeiro, a.3, n. 3, p 1-5, 2003

MORAES, R.M.; VINCENTE,E.;MORGANO, M.A.;MANTOVAVANI, D.M.B; BAIDINI, V.L.S. **Physico-chemical analyses of tilapia fillet.** São Paulo Agency for Agribusiness Technology. Institute of Food Technology, Analysis Reports CQ 7002/2004.

MORAIS, C.; MANTOVANI, D.M.B.; CARVALHO, C.R.L. Yield and Chemical composition of flesh ichthyofauna to catch shrimpseven- beards *(Xiphopenaeus kroyeri,* Heller, 1982). ***Coletâneas* do *ITAL*,** Campinas, v. 22, n.1, p. 62-72.

MPA - Ministry of Fisheries and Aquaculture. **Fisheries and Aquaculture Statistical Bulletin:** Brazil 2010. Brasília: MPA, 2012.129 p.

MUSTAFA, F.A.; MEDEIROS, D.M.Proximate composition, mineral content, and fatty acids of catfish *(Ictalaruspuntactus,* Rafinesque) for different seasons and cooking methods. **Journal of Food Science,** v.50, p.585-589,1985.

NAKATANIK; A. A.A.; BAUMGARTNER G.; BIALETZKI A.; SANCHES P.V.; CAVICCHIOLI-MAKRAKIS M.; PAVANELLIS C.S. **Ovos e larvas de peixes de água doce:** desenvolvimento e manual de identificação. Maringá: EDUEM/Nupélia, 2001.

NEPA/UNICAMP - Food Study and Research Centre of the State University of Campinas. **Brazilian table of food composition (TACO).4.ed.** rev. and ampl. Campinas: NEPAUNICAMP, 2011.161 P-

NOLLET, L.M.L. ; TOLDRÁ, F. **Handbook of seafood and seafood products analysis.** Boca Raton (USA): CRC Press - Taylor & Francis Group, 2010. 910p.

NOVAES, S. F.; CONTE-JUNIOR, C. A.; FRANCO, R. M.; MANO, S. B. Influence of new conservation technologies on food of animal origin. **Revista Científica Eletrónica de Medicina Veterinária,** Garça (SP),n.19, jul. 2012. 21 p.

NOVATO, P. F. C. **Comparison between demand, manual and automatic feeding systems on the performance of red tilapia *(Oreochromis spp).*** 2000. 87 p. Dissertation (Master's Degree in Aquaculture) - São Paulo State University, Jaboticabal, 2000.

NOVOA, D.Evaluación del efecto causado por la pesca de arrastre costera sobre la fauna íctica en la desembocadura del cano Manamo. **Acta Ecológica del Museo Marino de Margarita,** Caracas, n. 2, p. 43-62, 2000.

NUNES, E. S. C. L. **Evaluation of the physicochemical and microbiological quality of dried salted pirarucu *(Arapaíma gigas* Schinz, 1822) sold in the city of Belém, Pará.** 2011.77f. Thesis (Doctorate in Veterinary Hygiene Technological Processing of Products of Animal Origin), Universidade Federal Fluminense, Rio de Janeiro, 2011.

OETTERER, **M. Industrialisation of farmed fish.** Guaíba: Rditora Agropecuária, 2002. 200 p.

OETTERER, M. **Matéria-prima alimentaria:** pescado. São Caetano do Sul: Research Centre of the Mauá Institute of Technology, 1991.29p.

OFFER, G.; TRINICK, J. On the mechanism of water holding in meat: The swelling and shrinking of myofibrils. **Meat Science,** Barking, v .8, p. 245 - 281, 1983.

OFSTAD, R.; EGELANDSDAL, B.; KIDMAN, S.; OLSEN, R.; HERMANSSON, A. M Liquid loss as effected by post-mortem ultrastructural changes in fish muscle: cod *(Gadus Morhua* L.) and salmon *(Salmo salar).* **Journal of the Science of Food and Agriculture,** Sussex, v. 71, n. 3, p. 301-312, 1996 a.

OFSTAD, R.; KIDMAN, S.; MYKLEBUST, R.; OLSEN, R. L.; HERMANSSON, A. M. Liquid-holding capacity and structural changes in comminuted salmon *(Salmo salar).* Muscle as influenced by pH, salt and temperature. ***Lebensmittel-Wissenschaft***

und- Technologie, n.28, p. 329-339, 1995.

OFSTAD, R.; KIDMANB, S.; MYKLEBUSTC, R.; OLSEN, R.; HERMANSSONB, A. M. Factors influencing liquid-holding capacity and structural changes during heating of comminuted cod *(Gadus morhual L.)* muscle. ***Lebensmittel-Wissenschaft und-Technologie,*** v.9, n.1- 2, p 173 - 183, 1996b

OGAWA, M. e MAIA, E. L. Química do pescado: In: OGAWA, M. e MAIA, E. L. **Manual de pesca:** ciência e tecnologia do pescado. São Paulo: Livraria Varela, p. 29-71, 1999.

OGAWA, M. Fish as food. *In:* OGAWA, M., LIMA MAIA, E (Eds). **Fishing manual.** São Paulo: Varela, 1999. p. 03-05

OLIVEIRA, H.A.C.; GONÇALVES, A.A. Obstacle technology in fish products. *In:* **GONÇALVES, A.A.(Ed).Tecnologia do pescado:** ciência, tecnologia, inovação e legislação.São Paulo:Atheneu, 2011.p. 325 - 337.

OLIVEIRA, L. G.; PIANA, P. A.; LEMAINSKI, D.; BOMBARDELLI, R. A.; BOSCOLO, W. REIDEL, A. Carcass evaluation and morphometric characteristics of male and female Curimbatá (*Prochilodus lineatus)* and Piavuçú *(Leporinus macrochephalus). In:* CONGRESSO BRASILEIRO DE ENGENHARIA DE PESCA, 12., 2001, Foz do Iguaçú. Proceedings... Foz do Iguaçú: FAEP, 2001 (CD-ROM).

OLIVEIRA, R. L. M. **Morphometry, carcass yield and fillet composition of kissupirá *(Rachycentron canadum)* cultivated in net-tanks in the open sea off the coast of Pernambuco.** 2012. 64 f. Dissertation (Master's in Fisheries Resources and Aquaculture), Federal Rural University of Pernambuco, Recife, 2012.

OLIVO, R. Factors that influence the characteristics of meat raw materials and their technological implications. **Revista Nacional da Carne,** São Paulo, n. 307, p. 72-83, 2002.

OLSSON, G. B.; OLSEN, R. L.; OFSTAD, R. Post-mortem structural characteristics and water-holding capacity in Atlantic halibut muscle. **Lebensmittel-Wissenschat Und- Technologie - Food Science and Technology,** San Diego, v. 36, n. 1, p. 125-133, 2003.

ORDÓNEZ, J.A.; RODRÍGUEZ, M.I.C.; ÁLVAREZ, L.F.; SANZ, M.L.G.;

MINGUILLÓN, G.D.G.F.; PERALES, L.L.H.; CORTECERO,

M.D.S. **Food Technology:** Nutrient components and process. v.1, Porto Alegre: Editora Artmed 2005 b.

ORDÓNEZ, J.A.; RODRÍGUEZ, M.I.C.; ÁLVAREZ, L.F.; SANZ, M.L.G.; MINGUILLÓN, G.D.G.F.; PERALES, L.L.H.; CORTECERO, M.D.S. **Food Technology: Food of Animal Origin,** v.2. Porto Alegre: Artmed, 2005 a.

OSSA, P. Coronary heart disease and fish diet. **Boi. hosp.** San Juan de Dios,v. 32, n.5, p.34-4, 1995.

OSTRENSKY, A.; BORGHETTI, J. R.; PEDINI, M. Current situation of Brazilian and world aquaculture. *In:* VALENTI, V. C.; POLI, C. R.; PEREIRA, J. A.; BORGHETTI, J. R. **Aquaculture in Brazil:** bases for sustainable development. Brasília: Ministry of Science and Technology /CNPq. 2000. p. 353-382.

PADUA, D. M. **Fundamentos da Piscicultura.** Goiana: UCG Publishing House, 2000. 341 p.

PARSONS, T.; TAKAHASHI, M.; HARGRAVE, B. **Biological oceanographic processes.**3. ed. Pergamon Press: Oxford,1984. 330 p.

PEARSONS, A.M.; YOUNG, R.B. **Muscle and meat biochemistry.** Academic Press, Inc., San Diego. CA,1989

PEREIRA, T. J.F.; FERREIRA, L. K. S.; EVERTON, F. A.; FRAZÃO, F.B. Fishing in the Lençóis/Munin territory: fishing fleet and production levels. *In:* SEMANA NACIONAL DE OCEANOGRAFIA, 23,2011, São Luís, Anais...São Luís: UFMA, 2011. CD-ROM

PINHEIRO, R. S.B. SOBRINHO, A. G. S.; SOUZA, H. B. A.; YAMAMOTO, S. M.; POLIZEL NETO, A. Water retention capacity and water loss by cooking of sheep meat from different categories. *In:* CONGRESSO BRASILEIRO DE ZOOTECNIA, 19, 2009, Águas de Lindóia(SP), **Anais eletrónico...**Brasília: ABZ, 2009. Available at: < http://www.abz.org.br/publicacoes-tecnicas/anais-zootec/artigos-cientificos/tecnologia-produtos-origem-animal/3077-Capacidade-reteno-gua-

das- perdas-gua-por-coco-carne-ovinos-diferentes-categorias.html>. Accessed on:

POTTER, N. N.; HOTCHKISS, J. H. **Food science.** 5.ed. Zaragoza: Acribia, 1995. 667p.

POUEY, J.F.; STINGELIN, L.A. Carcass and meat yield of kingfish *(Odontesthes humensis)* in the 200 to 300g range. In: SIMPÓSIO BRASILEIRO DE AQUICULTURA, 9, 1996, Sete Lagoas, **Resumos...** Sete Lagoas: ABRAq, 1996. p.141.

REIDEL, A.; OLIVEIRA, L. G.; PIANA, P. A.; LEMAINSKI, D.; BOMBARDELLI, R. A.; BOSCOLO, W. R. Evaluation of yield and morphometric characteristics of male and female curimbatá *Prochilodus lineatus* (Valenciennes, 1836) and piavuçu *Leporinus macrocephalus* (Garavello & Britski, 1988). **Revista Varia Scientia,** Cascavel,v. 04, n. 08, p. 71-78, dec,2004

REIS, E. G. Reproduction and feeding habits of the marine catfish *Netuma barba* (Siluriformes, Ariidae) in the estuary of Lagoa os Patos, **Atlantic** Brazil. 1986. v. 8, p. 35-55.

REIS, R.E; KULLANDER, S.O.; FERRARIS JR., C.J. **Check list of the freshwater fishes of South and Central America.** Porto Alegre: EDIPUCRS. 2003. 729 p.

REIST, J. D. An empirical evaluation of several univariate methods that adjust for size variation in morphometric data. **Canadian Journal of Zoology,** Toronto, v. 63, n. 6, p. 1429-1439, 1985.

RIBEIRO, L.P.; LIMA, L.C.; TURRA E.M.; QUEIROZ, B.M.; RIBEIRO, T.G.; MIRANDA, M.O.T. Effect of weight and operator on fillet yield in red tilapia *Orechromis* spp. *In:* AQUICULTURA BRASIL, 98, 1998, Recife. **Proceedings...** Recife: ABRAq. 1998. v.2. p. 773-778.

RIOS, F. S.; DONATTI, L.; FERNANDES, M. N.; KALININ, A. L.; RANTIN, F.T. Effects of Food Deprivation in Muscle Structure and Composition of Traíra *(HopHas malabaricus):* Potential Implications on Flesh Quality. **Brazilian Archives of Biology and Technology,**v. 52, n.2, p. 465- 471,mar/abr, 2009.

RIOS, L. **Geography of Maranhão.** 4ª ed. rev. atual. São Luís: Central de Livros. 2005. 278p

ROÇA, R. O. **Properties of meat.** Botucatu(SP): UNESP, 2006. 10 p.

ROCHA, R.; FERRAZ, C.; LESSA, R.Morphometric characteristics of the flying fish *Cypselurus cyanopterus* caught in the region of the São Pedro and São Paulo Archipelago./n: JORNADA DE ENSINO, PESQUISA E EXTENSÃO DA UFRPE,9,2009, Recife, **Annuais eletrónicos...**Recife:UFRPE , 2009. Available at: www.eventosufrpe.com.br/jepex2009/cd/resumos/R1500-1 .p. Accessed on: 01 August 2012.

RODRIGUES, M. S. M.; RODRIGUES, L. B.;CARMO, J. L.; BRITO JÚNIOR, W.; PATEZ, C. Integral Utilisation of Fish with Emphasis on Hygiene, Handling, Cutting, Salting and Smoking. In: CONGRESSO BRASILEIRO DE EXTENSÃO UNIVERSITÁRIA, 02, 2004, Belo horizonte. **Electronic Proceedings...** BeloHorizonte :UFMG,2004.8p. Available at : http://www.ufmg.br/congrext/Tecno/Tecno7.pdf. Accessed on: 01 April 2012

ROMAN, J. A., SGARBIERE, V. C. Effect of enzymatic hydrolysis on the functional properties of bovine casein coagulated by the action of chymosin. **Food Science and Technology,** v.25, p. 468-474,2005.

RUSTAD, T. Peptides and proteins. *In:* NOLLET, L.M.L.; TOLDRÁ (Eds.). **Handbook of seafood and seafood products analysis.** BocaRaton: CRC Press-Taylor & Francis Group, 2010. p .11-20.

RUTTEN, M. J. M.; BOVENHUIS, H.; KOMEN, H. Genetic parameters for fillet traits and body measurements in Nile tilapia *(Oreochromis niloticus* L.). **Aquaculture,** v. 246, n. 1-4, p. 125-132, jan/feb, 2005.

RUTTEN, M.J.M. ; Bovenhuis, H.; Komen, H. Modelling fillet traits based on body measurements in three Nile tilapia strains *(Oreochromis niloticus).* **Aquaculture,** Amsterdam, v.231, n.1-4, p. 113-122, mar. 2004.

SÁ, E.M. F. The influence of water on meat properties. **Revista Nacional da Carne,** São Paulo, n.325, p.51-54, 2004.

SAKUMA A.M.et al. Sample collection. *In:* INSTITUTO ADOLFO LUTZ. **Physico-chemical methods for analysing food.** 4.ed.Brasília: ANVISA, 2005. p.44-54.

SANCHEZ, L. **Fish:** raw materials and processing. Campinas: Cargill Foundation, 1989.61p.

SANCHEZ, L.; GOMES, M. I; SASE, L. E. Storage of chilled Piauí hake, *Plagioscion squamosissimus* (HECKEL, 1840) i Evolution of chemical composition and some freshness indicators.Alim.Nutr., São Paulo, n.2, p.73-82, 1990

SANG N. V., THOMASSEN M., KLEMETSDAL G., GJOEN H. M. Prediction of fillet weight, fillet yield, and fillet fat for live river catfish *(Pangasianodon hypophthalmus).* **Aquaculture,**v.. 288,p. 166-171,2009.

SANTO, R.V.E.; ISAAC, V.J.; SILVA, L.M.A.; MARTINELLI, J.M.; HIGUCHI, H.; SAINT-PAUL, U. **Peixes e camarões do litoral bragantino.** Belém: Madam. 2005. 268 p.

SANTOS, A. B.; GINAR, R.M.B.; GALARÇA, R.C. G.; BRACCINI NETO, J. Fillet yield of palometa *(Serrasalmus spilopleura* Kner, 1860) in the Uruguaiana-RS/Brazil region. **Revista da FZVA,** Uruguaiana, v.13, n.2, p. 185-193, 2006

SANTOS, A.B., MELO, J.F.B.& LOPES, P.R.S. Study of the carcass of the rattlesnake *Hypostomus commersonii* in the Uruguaiana-RS/Brazil region. In: ENCONTRO SUL BRASILEIRO DE AQUICULTURA, 3; ENCONTRO RIOGRANDENSE DE TÉCNICOS EM AQUICULTURA, 6, 1995, Ibirubá. ***Proceedings...*** Porto Alegre: UFRS, 1995. p.70-76.

SANTOS, E. P. Referential allometry. **Boletim do Instituto de Pesca,** São Paulo, v. 21, p. 19 - 21, 1994. Unique.

SARMIENTO, A. M. L. **Microbiological, physico-chemical and sensory characteristics of tilapia fillets *(Oreochromis niloticus)* preserved in modified atmospheres under refrigeration.** 2006. 120 f. Dissertation (Master's Degree in Food Science and Technology), Federal University of Viçosa, Viçosa, 2006.

SERAFIM, C.F.S.; HAZIN, F. The coastal ecosystem. SERAFIM, C. F. S (Coord.); CHAVES, P. T.(Org.). **Geografia:** ensino fundamental e ensino médio: o mar no espaço geográfico brasileiro. Brasília: Ministry of Education, Secretariat of Basic Education, 2006. p. 101-131. (Coleção explorando o ensino , v.8).

SHEARER, K.D. Factors affeting the proximate composition of cultured fishes with emphasis on salmonids.Aquaculture, n.119, p.63-88, 1994.

SHIROTA, R.; SONADA,D.Y. Marketing fish in Brazil: characterising the markets. *In:* CYPRINO, J.E.P.; (Ed.); URBINATI, E.C.; PEREIRA, C.R. **Tópicos especiais em piscicultura de água doce tropical intensiva.** São Paulo: TecArt, 2004. p.501-516.

SIDONIO, L.; CAVALCANTI, I.; CAPANEMA, L.; MORCH, R.; MAGALHÃES,G.; LIMA, J.; BURNS,V.; ALVES JÚNIOR, A. J.; MUNGIOLI, R. Panorama of aquaculture in Brazil: challenges and opportunities, **BNDES Setorial,** n.35, p. 421 - 463, 2012. Available at: < http://www.bndes.gov.br/SiteBNDES/export/sites/default/bndes_pt/Galerias/Arquivos/conhecimento/bnset/set3512.pdf> . Accessed on: 17 September 2012.

SIIKORSKI, Z.E.; KOLAKOWSKA, A.; BUTT, J. R.Biochemical and microbial changes following capture. *In:* SIIKORSKI, **Z.E.(Eds).Tecnologia de lós productos del mar:** recursos,composicion nutricion y conservation. Zaragoza, Spain: Editorial Acribia, 1990.p. 73-101.

SIKORSKI, Z. E. **Tecnologia de los productos del mar: recursos, composicion nutritiva y conservación.** Zaragoza: Acribia, 1994. 330 p.

SIKORSKI, Z.E. **Seafood:** resources, nutritional, composition and preservation. Boca Raton(USA): CRC Press Inc,1990.248p.

SILVA, D. J.; QUEIROZ, A. C. **Food analysis:** chemical and biological methods. Viçosa: UFV, 2004. 233 p.

SILVA, E. G.; ALVES, E. L.; SANTOS, F. D. **Fish processing:** caranha filleting. 2008. 32f. Internship report (Agroindustry Technician), Technical School, Paraíso, 2008.

SILVA, V. R.C.E. **Physical and histological parameters of *octopus* species Octopus *vulgaris* and *Octopus insularis* submitted to conservation and cooking processes.** 2009.83f. Dissertation (Master's Degree in Food Science and Technology), Federal University of Ceará, Forteleza, 2009.

SMITH, N.J.H. **Fishing in the Amazon River.** Brasília: INPA/CNPq, 1979.154 P-

SOARES, M. **O oceano...nosso futuro: Relatório da Comissão Mundial Independente sobre os Oceanos.** Rio de Janeiro: CNIO, 1999. 250 p.

SOUZA, A.F.L.; INHAMUNS, A.J. Analysis of the meat yield of the main fish species commercialised in the State of Amazonas, Brazil. **Acta Amazônica,** Amazônia, v. 41, n. 2, p. 289 - 296, 2011.

SOUZA, H. B. A. Physical and sensory parameters used to evaluate the quality of chicken meat. *In:* SEMINÁRIO INTERNACIONAL DE AVES E SUÍNOS,5,2006, Florianópolis. Proceedings... Florianópolis: UNESP, 2006. p. 91-96.

SOUZA, M. L. R. **Processing of Nile tilapia *(Oreochromis niloticus)* fillet and skin*:*** technological aspects, centesimal composition, yield, shelf life of smoked fillet and resistance test of tanned skin. 2003. 169 f. Thesis (Doctorate in Zootechnics), Universidade Estadual Paulista, Jaboticabal, 2003.

SOUZA, M. L. R., LIMA, S., FURUYA, W.M., PINTO, A. A., LOURES, B. T. R. Sexrelated effects on the Processing yield of African catfish *(Ciarias gariepinus). In:* AQUICULTURA BRASIL,. Recife. **Anais** ...Recife, ABRAq ,1998 (CD-ROM).

SOUZA, M. L. R. Comparison of six filleting methods in relation to fillet yield and processing by-products of Nile tilapia *(Oreochromis niloticus).* **Revista Brasileira de Zootecnia,** Viçosa, v. 31, n.3, p. 1076-1054, 2002.

SOUZA, M. L. R; MACEDO-VIEGAS, E. M.; KRONKA, S. N. Influence of the filleting method and weight categories on the carcass yield of

Nile tilapia *(Oreochromis niloticus).* **Revista da Sociedade Brasileira de Zootecnia,** v. 28, n. 1, p. 1- 6, 1999.

SOUZA, **M.L.R. Technology for processing fish skins.**
Maringa: Eduem, 2004. 59p.(Fundamentum Collection:11)

SOUZA, M.L.R; MACEDO-VIEGAS, E.M. Effects of filleting methods on Processing yields of nile tilapia *(Oreochromis niloticus).* In: INTERNATIONAL SYMPOSIUM ON TILAPIA IN AQUACULTURE, 5,2000, Rio de Janeiro, **Electronic Annals...** Rio de Janeiro:ATA, p.415-457, 2000.

STANSBY, M.E. **Industrial fishery technology.** London: AVI, 1968. 393p.
STRIDE, R. K. **Diagnóstico da pesca Artesanal Marinha do Estado do Maranhão,** v.2. São Luís: CORSUP/EDUFMA, 1992. 205 p.

SU, K., HUANG, S., CHIU, C., SHEN, W. W.. Omega-3 fatty acids in major depressive disorder. A preliminary double-blind, placebo-controlled trial. **European Neuropsychopharmacology.** v. 13, p. 267-271, 2003.

SUDENE - Superintendence for the Development of the Northeast. **Research into the fishing resources of the Maranhão continental shelf.** Recife: Reprographics Division, 1976. 67p. (Fisheries Studies Series, 6).

SUZUKI, T. **Fish and Krill Protein:** Processing Technology. Applied Science Publishers. London, 1981.

SWAIN, D.P.; FOOTE, C.J. Stocks and chameleons: the use of phenotypic variation in stock identification. **Fish. Res.** n.43.p. 113-128, 1999.

SZPILMAN, M. **Peixes marinhos do Brasil:** guia prático de identificação. Rio de Janeiro: MAUAD Edirora, 2000. 288p.

TAVARES, M; GONÇALVES, A.A. Aspectos Físico-químicos do Pescado. In: **GONÇALVES, A.A.(Ed).Tecnologia do pescado:** ciência, tecnologia, inovação e legislação.São Paulo:Atheneu, 2011 .p. 10 - 20.

TAVARES,M.;MORENO, R.B.Fish and derivatives. *In:* INSTITUTO ADOLFO LUTZ. **Physical-chemical methods for analysing food.** 4Ed.Brasília: ANVISA, 2005. p.44-54.

TAYLOR, W.R. & MENEZES, N.A. Ariidae. In: FISCHER, W. (Ed.) **FAO species identification sheets for fishery purposes.** West Atlantic (Fishing Area 31). v. 1. Rome: FAO, 1978.

TELLES, A. C.; BAGATINI, D. C.; PIZATO, S.; CORTEZ-VEGA, W. R.; PRENTICE-HERNANDEZ, C. Physico-chemical evaluation of rocket hake fillets *(Macrodon ancylodon)* packaged in normal atmosphere.In: MOSTRA DE PRODUÇÃO UNIVERSITÁRIA , 9, 2010, Rio Grande(RS), **Anais eletrónicos...**Rio Grande: FURG, 2010. Available at: < 200.132.208.55/anaismpu/cd2010/cic/227.doc>. Accessed on : 04 September 2012

TORNBERG, E. Effects of heat on meat proteins. Implications on structure and quality of meat products. **Meat Science,** Oxford, v. 70, n. 3, p. 493-508,2005.

TRONDSEN, T., BRAATEN, T., LUND, E., EGGEN, A. E. Consumption of seafood -

the influence of overweight and health beliefs. **Food Quality and Preference,** Oxford, v. 15, n. 4, p. 361-374, 2004.

TSCHEUSCHNER, H. D. **Fundamentos de tecnologia de los alimentos.** Zaragoza: Acribia, 2001.746p.

VALENTI, W.C.; POLI, C.R.; PEREIRA, J.A. Aquaculture in Brazil: bases for sustainable development. Brasília: Ministry of Science and Technology, 2000. 399p.

VENTURINI, K. S.; SARCINELLI, M. F.; SILVA, L. C. **Characteristics of chicken meat.** Espírito Santo: UFES, 2007, 7 p. (Technical Bulletin - PIE- UFES:01307).

VILAS BOAS, G. C. **Morphometry, processing yield and chemical composition of matrinchã fillet *Brycon cephalus* (GUNTHER, 1869).** 2001. 59 p. Dissertation (Master's Degree in Zootechnics) - Federal University of Lavras, Lavras, MG.

WAITZBERG, D.; GALIZIA, M.; HORIE, L. Carbohydrates. *In:* WAITZBERG, D.L.(Ed.). **Oral, enteral and parenteral nutrition in clinical practice.** 4.ed.São Paulo: Atheneu, 2009. p 55-84.

WILLE, K.; MCLEAN, E., GODDARD, J.S.; BYATT, J.C. Dietary lipid levei and growth hormone alter growth and body conformation of blue tilapia *Oreochromis aureus.* **Aquaculture,**n. 209,p. 219-232, 2002.

WOSIACKI, W. B.; OYAKAWA, O. T.Two new species of the catfish genus *Trichomycterus* (Siluriformes: Trichomycteridae) from the rio Ribeira de Iguape Basin, Southeastern BraziL **Neotropical Ichthyology,** n. 3, n.4, p.465-472, 2005.

XAVIER, A. A. S. **Development and characterisation of piranha sausage *(Serrasalmus* sp.).2009.** 88f. Dissertation (Master's in Food Science and Technology), Federal University of Ceará, Fortaleza, 2009.

ZAITSEV, V.; KIZEVETTER, I.; LAGUNOV, L.; MAKAROVA, T.; MINDER, L.; PODSEVALOV, V. **Fish Curing and Processing.** Honolulu/Hawaii: University Press of the Pacific, 2004

CHAPTER 8

APPENDICES

Appendix A. Morphometric characteristics and relationships of the uritinga specimens *(Sc/acfs proops)*

SAMPLE	CT[1] (cm)	CP[2] (cm)	CZ[3] (cm)	LD[4] (cm)	CC[5] (cm)
01	41,00	34,00	35,00	5,20	8,70
02	39,50	32,00	33,00	5,00	7,21
03	39,50	30,00	33,00	5,30	7,54
04	38,00	29,50	33,20	5,55	8,00
05	41,30	31,50	35,40	5,20	8,50
06	35,00	27,40	31,50	4,83	6,84
07	36,70	28,90	32,00	4,90	8,60
08	42,50	35,00	38,00	5,63	9,00
09	45,00	37,00	38,50	6,44	9,64
10	41,50	32,00	35,50	5,10	7,85
11	39,50	31,50	33,50	4,95	7,55
12	41,00	31,00	34,00	5,53	9,00
13	44,00	36,50	39,00	4,72	9,30
14	39,20	31,00	33,50	5,24	8,83
15	42,50	36,50	38,00	6,10	9,30
16	41,50	34,00	35,50	5,80	8,90
17	41,00	34,50	35,50	6,10	9,00
18	46,00	38,50	40,00	6,10	9,80
19	41,50	34,50	36,00	5,70	9,10
20	44,50	37,00	39,00	5,94	9,20
21	42,50	36,00	37,00	5,94	9,10
22	40,00	33,50	35,00	5,20	9,10
23	40,00	34,50	35,50	4,50	7,80
24	41,00	35,00	36,50	4,90	8,10
25	40,50	34,50	36,00	5,05	7,80
26	45,00	36,50	39,00	5,64	9,20
27	41,50	34,00	36,00	5,00	7,80
28	40,50	33,50	36,00	5,60	7,80
29	40,00	34,00	36,00	5,22	7,80
30	42,50	35,50	37,50	5,63	9,10

[1] Total Length,[2] Standard Length,[3] Zoological Length,[4] Dorsal Height,[5] Head Length

SAMPLE	CC /LD[12] (%)	CC/CP[3] (%)	LD/CP (%)
01	167,31	25,59	15,29
02	144,20	22,53	15,63
03	142,26	25,13	17,67
04	144,14	27,12	18,81
05	163,46	26,98	16,51
06	141,61	24,96	17,63
07	175,51	29,76	16,96
08	159,86	25,71	16,09
09	149,69	26,05	17,41
10	153,92	24,53	15,94
11	152,53	23,97	15,71
12	162,75	29,03	17,84
13	197,03	25,48	12,93
14	168,51	28,48	16,90

15	**152,46**	**25,48**	**16,71**
16	**153,45**	**26,18**	**17,06**
17	**147,54**	**26,09**	**17,68**
18	**160,66**	**25,45**	**15,84**
19	**159,65**	**26,38**	**16,52**
20	**154,88**	**24,86**	**16,05**
21	**153,20**	**25,28**	**16,50**
22	**175,00**	**27,16**	**15,52**
23	**173,33**	**22,61**	**13,04**
24	**165,31**	**23,14**	**14,00**
25	**154,46**	**22,61**	**14,64**
26	**163,12**	**25,21**	**15,45**
27	**156,00**	**22,94**	**14,71**
28	**139,29**	**23,28**	**16,72**
29	**149,43**	**22,94**	**15,35**
30	**161,63**	**25,63**	**15,86**

[1] Head length,[2] Dorsal height;[3] Standard length

Appendix B. Descriptive statistics of the characteristics and morphometric relationships of the *S.proops* specimens

VARIABLES	STATISTICAL PARAMETERS				
	MAXIMUM VALUE	MINIMUM VALUE	AVERAGE	STANDARD DEVIATION	COEFFICIENT OF VARIATION
CT[1] (cm)	**46,00**	**35,00**	**41,14**	**2,38**	**5,79%**
CP[2] (cm)	**38,50**	**27,40**	**33,64**	**2,67**	**7,95%**
CZ[3] (cm)	**40,00**	**31,50**	**35,79**	**2,21**	**6,17%**
LD[5] (cm)	**6,44**	**4,50**	**5,40**	**0,48**	**8,87%**
CC[6] (cm)	**9,80**	**6,84**	**8,52**	**0,77**	**9,06%**
CC/LD (%)	**197,03**	**139,29**	**158,07**	**12,26**	**7,75%**
CC/CP (%)	**29,76**	**22,53**	**25,35**	**1,87**	**7,39%**
LD/CP (%)	**18,81**	**12,93**	**16,10**	**1,36**	**8,42%**

[1] Total length ;
[2] Standard length
[3] Zoo length
[4] Dorsal height
[5] Head length

Appendix C. Fillet yields of uritinga specimens *(Sciads proops)*

SAMPLE	Pi[1] (g)	skin-on fillet weight (g)			SKIN-ON FILLET YIELD (%)		
		TOTAL WEIGHT	FCPE[2]	FCPD[3]	TOTAL	FCPE	FCPD
01	515,70	242,70	130,30	112,40	47,06	25,27	21,80
02	440,40	213,10	105,10	108,00	48,39	23,86	24,52
03	413,90	201,80	102,40	99,40	48,76	24,74	24,02
04	426,20	172,20	76,80	95,40	40,40	18,02	22,38
05	547,30	279,40	145,30	134,10	51,05	26,55	24,50
06	318,80	146,80	77,70	69,10	46,05	24,37	21,68
07	356,70	142,30	69,50	72,80	39,89	19,48	20,41
08	596,70	285,80	140,90	144,90	47,90	23,61	24,28
09	718,80	343,40	179,90	163,50	47,77	25,03	22,75
10	517,30	261,80	128,90	132,90	50,61	24,92	25,69
11	441,50	222,30	112,80	109,50	50,35	25,55	24,80
12	542,80	263,60	131,70	131,90	48,56	24,26	24,30

13	665,90	348,70	174,80	173,90	52,37	26,25	26,12
14	465,20	230,30	124,60	105,70	49,51	26,78	22,72
15	641,10	316,80	157,80	159,00	49,42	24,61	24,80
16	571,20	280,60	140,30	140,30	49,12	24,56	24,56
17	522,70	249,90	125,90	124,00	47,81	24,09	23,72
18	745,30	376,00	189,60	186,40	50,45	25,44	25,01
19	578,30	302,70	155,30	147,40	52,34	26,85	25,49
20	694,10	371,30	184,30	187,00	53,49	26,55	26,94
21	625,10	324,50	166,70	157,80	51,91	26,67	25,24
22	518,00	243,10	122,40	120,70	46,93	23,63	23,30
23	524,00	268,20	142,00	126,20	51,18	27,10	24,08
24	490,70	234,30	130,00	104,30	47,75	26,49	21,26
25	510,00	249,10	128,00	121,10	48,84	25,10	23,75
26	689,30	344,10	186,10	158,00	49,92	27,00	22,92
27	472,70	229,80	114,50	115,30	48,61	24,22	24,39
28	515,50	255,80	124,90	130,90	49,62	24,23	25,39
29	534,20	279,10	141,70	137,40	52,25	26,53	25,72
30	635,80	337,30	172,40	164,90	53,05	27,12	25,94

[1] Weight of whole fish;[2] Fillet with skin on left side;[3] Fillet with skin on right side.

SAMPLE	FILE WEIGHT WITHOUT SKIN (g)			SKINLESS FILLET YIELD (%)		
	TOTAL WEIGHT	FSPE[1]	FSPE[2]	TOTAL	FSPE	FSPD
01	186,80	90,50	96,30	36,22	17,55	18,67
02	170,60	86,40	84,20	38,74	19,62	19,12
03	168,60	83,60	85,00	40,73	20,20	20,54
04	154,10	72,00	82,10	36,16	16,89	19,26
05	238,10	122,30	115,80	43,50	22,35	21,16
06	117,40	61,60	55,80	36,83	19,32	17,50
07	117,90	58,30	59,60	33,05	16,34	16,71
08	240,20	114,80	125,40	40,25	19,24	21,02
09	281,40	140,80	140,60	39,15	19,59	19,56
10	218,80	108,70	110,10	42,30	21,01	21,28
11	183,00	95,60	87,40	41,45	21,65	19,80
12	216,20	109,60	106,60	39,83	20,19	19,64
13	286,10	143,50	142,60	42,96	21,55	21,41
14	192,90	103,30	89,60	41,47	22,21	19,26
15	255,20	124,50	130,70	39,81	19,42	20,39
16	228,50	108,10	120,40	40,00	18,93	21,08
17	199,50	95,80	103,70	38,17	18,33	19,84
18	305,40	154,80	150,60	40,98	20,77	20,21
19	258,20	132,50	125,70	44,65	22,91	21,74
20	320,30	162,60	157,70	46,15	23,43	22,72
21	269,70	138,70	131,00	43,15	22,19	20,96
22	197,70	101,80	95,90	38,17	19,65	18,51
23	222,20	116,60	105,60	42,40	22,25	20,15
24	196,40	104,40	92,00	40,02	21,28	18,75
25	204,70	104,70	100,00	40,14	20,53	19,61
26	277,30	147,70	129,60	40,23	21,43	18,80
27	186,60	90,20	96,40	39,48	19,08	20,39
28	215,40	106,10	109,30	41,78	20,58	21,20
29	233,70	118,10	115,60	43,75	22,11	21,64
30	284,00	143,50	140,50	44,67	22,57	22,10

[1]Skinless fillet on the left;[2] Skinless fillet on the right

Appendix D. Descriptive statistics for fillet yields of uritinga specimens

(Sc/acfs proops)

VARIABLES	STATISTICAL PARAMETERS				
	MAXIMUM VALUE	MINIMUM VALUE	AVERAGE	STANDARD DEVIATION	COEFFICIENT OF VARIATION
Pi'(g)	745,30	318,80	541,17	104,31	19,28
WEIGHT OF FILLET WITH SKIN					
Total weight (g)	376,00	142,30	267,23	61,01	22,83
FCPE² (g)	189,60	69,50	136,09	31,95	23,48
FCPD³ (g)	187,00	69,10	131,14	29,82	22,74
FILLET YIELD WITH SKIN					
Total yield (%)	53,49	39,89	49,05	3,08	6,28
FCPE (%)	27,12	18,02	24,96	2,05	8,20
FCPD (%)	26,94	20,41	24,08	1,55	6,45
WEIGHT OF SKINLESS FILLET					
Total weight (g)	320,30	117,40	220,90	51,01	23,09
FSPE⁴ (g)	162,60	58,30	111,37	26,42	23,72
FSPD⁵ (g)	157,70	55,80	109,53	25,12	23,93
SKINLESS FILLET YIELD '					
Total yield (%)	46,15	33,05	40,54	2,84	7,01
FSPD (%)	22,91	16,34	20,42	1,75	8,57
FSPE (%)	23,43	16,71	20,12	1,41	7,03

[1] Weight of whole fish;[2] Fillet with skin on left side;[3] Fillet with skin on right side;[4] Fillet without skin on left side;[5] Fillet without skin on right side

Appendix E. Carcass and waste yields of uritinga specimens *(Sc/ads proops)*

SAMPLE	CARcass WEIGHT (g)		CARCASS YIELD (%)	
	WITH A HEAD	WITHOUT A HEAD	WITH A HEAD	WITHOUT A HEAD
01	476,60	308,10	92,42	59,74
02	410,40	259,60	93,19	58,95
03	382,00	247,00	92,29	59,68
04	388,80	220,50	91,22	51,74
05	508,90	340,00	92,98	62,12
06	289,10	185,40	90,68	58,16
07	324,30	187,30	90,92	52,51
08	545,40	350,90	91,40	58,81
09	654,00	401,10	90,98	55,80
10	479,90	316,80	92,77	61,24
11	410,00	263,10	92,87	59,59
12	499,40	309,90	92,00	57,09
13	633,30	407,30	95,10	61,17
14	432,00	277,70	92,86	59,69
15	578,00	368,40	90,16	57,46
16	528,10	334,60	92,45	58,58
17	487,80	304,30	93,32	58,22
18	661,40	429,50	88,74	57,63
19	515,60	335,10	89,16	57,95
20	649,70	440,80	93,60	63,51
21	580,60	377,50	92,88	60,39
22	484,90	311,10	93,61	60,06
23	488,10	326,00	93,15	62,21
24	453,60	288,50	92,44	58,79
25	474,20	303,50	92,98	59,51
26	623,30	408,60	90,43	59,28

27	431,10	268,90	91,20	56,89
28	481,00	310,10	93,31	60,16
29	493,10	329,80	92,31	61,74
30	595,30	403,40	93,63	63,45

FERRE1RA, L. K. S. - APEND1CE AND

SAMPLE	WASTE WEIGHT (g)			WASTE YIELD (%)		
	HEAD	VISCERAS	TOTAL WASTE	HEAD	VISCERAS	TOTAL WASTE
01	168,50	37,90	206,40	32,67	7,95	40,02
02	150,80	30,50	181,30	34,24	7,43	41,17
03	135,00	30,00	165,00	32,62	7,85	39,86
04	168,30	36,30	204,60	39,49	9,34	48,01
05	168,90	36,50	205,40	30,86	7,17	37,53
06	103,70	28,10	131,80	32,53	9,72	41,34
07	137,00	31,40	168,40	38,41	9,68	47,21
08	194,50	45,30	239,80	32,60	8,31	40,19
09	252,90	57,10	310,00	35,18	8,73	43,13
10	163,10	35,50	198,60	31,53	7,40	38,39
11	146,90	28,90	175,80	33,27	7,05	39,82
12	189,50	40,50	230,00	34,91	8,11	42,37
13	226,00	44,80	270,80	33,94	7,07	40,67
14	154,30	32,00	186,30	33,17	7,41	40,05
15	209,60	60,50	270,10	32,69	10,47	42,13
16	193,50	39,90	233,40	33,88	7,56	40,86
17	183,50	34,20	217,70	35,11	7,01	41,65
18	231,90	62,40	294,30	31,11	9,43	39,49
19	180,50	38,90	219,40	31,21	7,54	37,94
20	208,90	48,10	257,00	30,10	7,40	37,03
21	203,10	44,30	247,40	32,49	7,63	39,58
22	173,80	39,20	213,00	33,55	8,08	41,12
23	162,10	35,60	197,70	30,94	7,29	37,73
24	165,10	41,70	206,80	33,65	9,19	42,14
25	170,70	34,40	205,10	33,47	7,25	40,22
26	214,70	54,70	269,40	31,15	8,78	39,08
27	162,20	33,60	195,80	34,31	7,79	41,42
28	170,90	42,70	213,60	33,15	8,88	41,44
29	163,30	38,40	201,70	30,57	7,79	37,76
30	191,90	38,80	230,70	30,18	6,52	36,28

Appendix F. Descriptive statistics for carcass and waste yields of uritinga specimens *(Sc/acfs proops)*

VARIABLES	STATISTICAL PARAMETERS				
	MAXIMUM VALUE	MINIMUM VALUE	AVERAGE	STANDARD DEVIATION	COEFFICIENT OF VARIATION (%)
CARCASS WEIGHT					
With head	661,40	289,10	498,66	95,01	19,05%
Headless	440,80	185,40	320,49	66,27	20,68%
CARCASS YIELD					
With head	95,10	88,74	92,17	1,42	1,54%
Headless	63,51	51,74	59,07	2,65	4,49%
WEIGHT OF WASTE					
Head	252,90	103,70	178,17	31,31	17,58%

Guts	62,40	28,10	40,07	9,00	22,47%
Total waste	310,00	131,80	218,24	39,40	18,05%
WASTE YIELD					
Head	39,49	30,10	33,10	2,15	6,50%
Guts	10,47	6,52	8,06	0,98	12,14%
Total waste	48,01	36,28	40,52	2,57	6,35%

Appendix G. Physico-chemical parameters of uritinga fillets *(Sciads proops)*

SAMPLE	UM1(%)	PB² (%)	pH³	WATER LOSS		CRA
				DRIP LOSS (%)	*COOKING LOSS (%)*	PRESSURE (%)
01	78,14	18,27	7,10	3,06	20,47	46,39
02	78,34	17,73	7,00	4,71	18,38	52,79
03	78,54	18,62	7,10	6,31	23,34	50,14
04	78,54	18,94	7,00	7,31	16,75	54,82
05	78,47	12,82	7,60	6,83	22,46	51,01
06	78,46	19,59	7,60	5,03	5,21	55,11
07	78,64	18,96	7,70	4,17	12,70	52,28
08	78,56	18,04	7,70	6,28	35,05	49,86
09	78,44	16,46	7,60	5,48	38,35	-
10	78,34	17,95	7,60	4,93	19,48	-

[1] **Humidity**

[2] **Crude Protein**

[3] **Hydrogen potential**

Appendix H. Descriptive statistics for the physico-chemical parameters of uritinga fillets *(Sc/ads proops)*

VARIABLES	STATISTICAL PARAMETERS				
	MAXIMUM VALUE	MINIMUM VALUE	AVERAG E	STANDARD DEVIATION	COEFFICIENT OF VARIATION (%)
Humidity (%)	78,64	78,14	78,45	0,14	0,18%
Crude Protein (%)	19,59	12,82	17,74	1,93	10,86%
PH	7,70	7,00	7,40	0,31	4,13%
Water loss					
Drip loss (%)	7,31	3,06	5,41	1,30	0,24
Cooking loss (%)	38,35	5,21	21,22	9,72	0,46
CRA					
Pressure (%)	55,11	46,39	51,55	2,86	6,16%

[1] Total length

[4] Dorsal height

MIX
Papier aus verantwortungsvollen Quellen
Paper from responsible sources
FSC® C105338

Printed by Books on Demand GmbH, Norderstedt / Germany